Amazonia

WORLDS OF MAN

Studies in Cultural Ecology

EDITED BY
Walter Goldschmidt

University of California
Los Angeles

Amazonia

Man and Culture in a Counterfeit Paradise

BETTY J. MEGGERS
SMITHSONIAN INSTITUTION

ALDINE • ATHERTON | CHICAGO • NEW YORK

The Author

Betty J. Meggers earned her A.B. from the University of Pennsylvania, M.A. from the University of Michigan, and Ph.D. from Columbia University. Dr. Meggers is currently Research Associate, Department of Anthropology, National Museum of Natural History, Smithsonian Institution. She has taught at American University, served as Executive Secretary of the American Anthropological Association, and has been a consultant to Battelle Memorial Institute. She has done extensive field research and is a noted authority on South America, especially on the archeology of the west coast and, more generally, on the cultures of the lowlands. She has been awarded the Washington Academy of Sciences Award for Scientific Achievement, the Gold Medal of the 37th International Congress of Americanists, and the Order of Merit from the Government of Ecuador. Her contributions to the literature include more than 100 articles, book reviews, and translations

First published 1971 by
Aldine · Atherton, Inc.
529 South Wabash Avenue
Chicago, Illinois 60605

Library of Congress Catalog Number 74–141427
ISBN 202–01016–3, cloth; 202–01044–9, paper

Designed by John Goetz
Printed in the United States of America

Foreword

The tropical rain forest retains for modern man a peculiar mystique: dark and foreboding and filled with primordial fears and at the same time rich and colorful as painted by Henri Rousseau or described by W. H. Hudson in *Green Mansions.* It has been more resistant to the invasions of modern technology than even the arctic, and stands as a constant challenge to those who feel that nature should be subordinated to the will of man and who see in the rich carpet of the forest the potentialities of great abundance.

In *Amazonia,* Betty Meggers examines this landscape and the uses to which it has been put by the Indians who have made it their home from time immemorial. She demonstrates how native peoples have learned to exploit this environment which, while not foreboding, is certainly not rich in human resources. She brings to this analysis an ecological approach.

Ecology began as the study of the interaction between one living entity—usually a species—and all other living and inanimate elements in its environment. It is a demanding exercise, for the operative word is interaction—the force exerted by the species upon others as well as the influence these others have upon the species under investigation. Quite clearly it bears only superficial resemblance to naive environmentalism, and properly used it is a powerful concept for explaining the diverse forms of life and the adaptive processes of evolution.

Cultural ecology borrows this concept, replacing the species with a social unit—usually the tribe or community—and recognizing that the social behavior patterns we call culture are similarly adaptive to the environment through dynamic interaction. This is not merely a matter of seeking sustenance and shelter; it involves the collaborative actions that we call social organization and the patterns of belief we call religion; indeed the whole range of cultural behavior. *Worlds of Man* is a series dedicated to the exploration of these relationships in diverse settings and circumstances.

Dr. Meggers' analysis of five living cultures in the great area the Brazilians call *terre firme* and of two extinct cultures in the much smaller and richer area they call the *várzea,* shows us some of the basic details of this adaptive process. She gives us a means of understanding the functional importance of institutions such as warfare and infanticide, among other collaborative social actions. Man, like other animals, is thus seen as an integral part of nature, neither as its conqueror nor its slave, and significantly, the native populations recognize this harmony.

The practical problems Dr. Meggers addresses in her brief postscript relate to modern man's uses of this environment. She shows us that the exploitative techniques appropriate to the temperate homeland of Western civilization are not merely unworkable in this tropical environment, with its intense sunshine and torrential rains, but actually destroy the resources themselves. Some of these maladaptations lie in purely technical matters, such as the destruction of soil fertility by exposure of the land to the tropical sun and rain or the keeping of cattle under adverse conditions. Others, however, are institutional in character, and in her concluding remarks Dr. Meggers points out how we must bend institutions to the requirements of the land, rather than endeavoring to bend the landscape to our customary ways of life. There is no better example of the practical implications inherent in the thesis of a cultural ecology.

Walter Goldschmidt

Preface

An anthropologist working in Amazonia finds himself vastly outnumbered by representatives of other disciplines. Zoologists, botanists, geologists, limnologists, and other scientists have begun to compensate for decades of neglect by escalating research. The absence of a similar upsurge of interest among anthropologists is particularly unfortunate since the aboriginal population is rapidly changing in the face of encroaching civilization. In 1957, when Darcy Ribeiro warned that "It is fully certain that there will never again be an opportunity to do whatever is not done now," there were 143 indigenous groups remaining in Brazil, two-thirds of them in Amazonia. Since that date, acculturation and extinction have accelerated, yet the warning goes unheeded.

There are two reasons for being concerned. One is that the advancement of anthropology as a scientific discipline depends on the ability to test hypotheses against a broad spectrum of data. Aboriginal cultures are not only less viable than geological formations, animals, and plants, but they can change at a much faster rate, so that time is not on our side. The hiatus that exists in our body of ethnographic data on Amazonian groups will be permanent if steps to fill it are not taken at once.

These remaining societies have a significance beyond their value to anthropological theory. To the extent that they constitute mature adaptations to a particular type of environment, they provide us with a perspective on an ecosystem as a whole that is different from that obtainable from any other kind of scientific investigation. Where the breakdown of the equilibrium adaptation has already begun, research can further our understanding of its effects on the environment and on the people. Anthropological data can no longer be viewed as curious facts of no practical use; anthropologists have something essential to contribute and an obligation to do so. If we persist in ignoring our mission, not only Amazonia but the entire planet may become an unsuitable habitat for man.

In preparing this book, I have drawn upon more than twenty years of experience in the tropical lowlands of South America, with the result that it is impossible to mention all of the individuals who have contributed to the formulation of the ideas that are expressed. A number of people have generously provided data or illustrations, however, and I should like to acknowledge their cooperation with grateful thanks: Protasio Frikel, who clarified aspects of the Xikrín (Kayapó) subsistence practices; Hilgard O'Reilly Sternberg, who made available his unpublished study of the várzea near Manaus; Kalervo Oberg, who supplied photographs of the Camayurá; C. R. Jones, who provided several of the Waiwai; Jack Marquardt, who tracked down many obscure references; Harald Sioli and Clifford Evans, who served as preliminary sounding boards for some of the interpretations. Photographs of the Jívaro were furnished by National Anthropological Archives of the Smithsonian Institution. Maps and diagrams were drawn by George Robert Lewis.

Contents

INTRODUCTION

Man's biological unity with other organisms, long a matter of passionate dispute, is rarely contested today. Scientific research has made it abundantly clear that all living things possess the same basic structure and composition and has revealed many of the circumstances responsible for evolutionary differentiation. Once this biological relationship has been acknowledged, however, it is often dismissed as irrelevant to an understanding of cultural development, on the assumption that culture is immune to natural selection.

When the human situation of half a million years ago is compared with the present, the contrast is impressive. Although the adaptive impact of culture was small for many millennia, gains have been cumulative. Simultaneously, the rate of change, formerly so slow that it was invisible to several generations of observers, has accelerated to the point that our surroundings are transformed before our eyes. The capacity to protect ourselves from starvation, disease, and death from injury is so great that population proliferation has become a cause for alarm. We have remodeled the surface of the planet to our liking, bringing water to deserts, draining swamps and lakes, leveling hills, and altering the natural vegetation to a degree formerly matched only by major geological or climatic transformations, most of which proceeded at a significantly slower rate. We hold the power of life and death over our fellow vertebrates and have already pushed dozens of species into extinction. As we look about us, we do not see a delicately balanced ecosystem in which man is an increasingly discordant element; instead, we see our own egosystem, where all forms of life except human beings and a few domesticated plants and animals are either undesirable or superfluous.

Men have not always taken this view. Primitive peoples regard themselves as part of nature, neither superior nor inferior to other creatures (although often superior to other groups of men). The souls of human beings are believed to be capable of entering the bodies of animals and vice versa, and animal spirits are often thought to exercise significant control over human destiny. Such supernatural concepts are a translation into cultural terms of checks and balances that exist on a biological level to maintain the equilibrium of an ecosystem. Since unrestricted use of increasingly efficient hunting techniques would cause rapid exhaustion of the very resource they were designed to make more accessible, supernatural sanctions develop to limit or channel their use. This kind of functional relationship between a religious belief and a tool is one example of the infinite number of links that serve both to bind a cultural system together and to make it a compatible part of an ecosystem as a whole.

The existence of a common denominator for biological and cultural phenomena is implied by their parallel evolutionary trends. The elaboration of unicellular organisms into higher mammals has its counterpart in the transformation of hunting bands into urban nations. For billions of years, organisms remained simple, small, and little changed; similarly, for thousands of years cultures exhibited no significant growth in complexity. Once differentiation began, however, it proceeded at an accelerated pace on both levels. Organisms and cultures developed new characteristics that permitted them to invade previously inaccessible habitats or to exploit the old ones in novel ways. The paleontological and archeological records both testify to the fact that many adaptive pathways were blind alleys, ultimately leading to extinction, while others branched into new and unexpected directions. The culmination of this process is the extraordinarily diversified and intricately integrated biosphere that makes our planet unique in the solar system and, possibly, in the universe.

The attempt to understand man's place in this biosphere is hampered by a severe complication: our inability to look at ourselves without bias. Culture always screens or slants our perceptions because our own ideas, attitudes, beliefs, and thought processes are integrally linked with the phenomena we wish to study. Anthropologists have tried to surmount this handicap by focusing on primitive cultures, which are not only simpler than our own and presumably easier to unravel, but are also alien so that the line separating the observer from the observed is easier to draw. While human emotions nearly always intrude, this approach has met with with considerable success.

Judged by the criterion of novelty, Amazonia is an ideal place for temperate zone investigators to study cultural adaptation. The riverine labyrinth and trackless forest hold a mystery for us that has lost little of its enchantment during the centuries since their discovery by European explorers. Even today, when the challenge of outer space makes most other terrestrial regions seem commonplace, Amazonian statistics remain impressive. With a flow five times that of the Congo and 12 times that of the Mississippi, the Amazon contributes nearly one-fifth of all the water annually received into the oceans. It disgorges as much water into the Atlantic every 24 hours as the Thames carries past London in a year. This output is even more remarkable in view of the insignificant slope of the basin; the peak of the Empire State building is four times the elevation of Iquitos in eastern Peru, which lies 2,300 miles from the mouth of the Amazon. Soundings on Marajó Island have revealed sediment accumulation to a depth of 12,600 feet, almost as far below sea level as the elevation above sea level of La Paz in the highlands of Bolivia. While the features figuring in these world records are not the most significant for human adaptation, their spectacular character conveys an accurate impression of the uniqueness of the tropical lowland environment of South America.

Amazonia is an appropriate laboratory for examining cultural adaptation for another reason. During the last few millennia, it has been exposed to two successive and distinct kinds of human utilization. The first was evolved under the influence of natural selection from the ingredients brought by the initial human settlers several millennia before the Christian era. The second, introduced at the opening of the sixteenth century, was an externally controlled exploitative system that not only destroyed the previous equilibrium but prevented establishment of a new balance. The examination of these two markedly contrasting kinds of exploitation of the same environment permits recognition of significant aspects of the culture-environment relationship that might otherwise remain obscure.

During the ensuing examination of man in the context of the Amazonian ecosystem, two propositions will be accepted as valid: 1) man is an animal and, like all other animals, must maintain an adaptive relationship with his surroundings in order to survive; and 2) although he achieves this adaptation principally through the medium of culture, the process is guided by the same rules of natural selection that govern biological adaptation. Examination of the interaction of culture and environment requires a compilation of the significant facts in both categories of phenomena. This is not as formidable a task as it might at first appear, however, because no organism interacts equally intimately with all aspects of its environment. Since subsistence is a primary requirement for life, those characteristics of soil, topography, climate, flora, and fauna most relevant to the quality and quantity of the food supply can be selected for emphasis. Approaching Amazonia from this point of view permits recognition of two subregions of markedly contrasting size and differing subsistence potential: 1) the vast terra firme, where resources are thinly dispersed but continuously available; and 2) the narrow floodplain, or várzea, where scarcity alternates with abundance as the river rises and falls. If adaptation is a primary determinant of culture, we should expect to find differences in the cultural complexes associated with these two subregions. This is indeed the case.

The recognition of cultural differences that have adaptive significance is facilitated by the existence of comparable descriptions of the principal features of a representative sample of cultures. Five groups have been chosen to illustrate the range of variation on the terra firme, and two more to exemplify adaptation to the várzea. The similarities and differences that exist between these cultural complexes reveal a great deal about the intensity of environmental constraints and the flexibility of cultural response. The interplay between culture and environment that emerges from this analysis provides a basis for formulating additional hypotheses about the general process of cultural evolution.

The story does not end at this point, however. Amazonia today is a very different place than it was before A.D. 1500—not because the climate or topography has altered appreciably but because the cultural increment has drastically changed. The degradation that has taken place in the habitat, particularly during the past half-century, provides a clear demonstration of the environment-culture relationship in its most disharmonious form. The persistence of the myth of boundless productivity in spite of the ignominious failure of every large-scale effort to develop the region constitutes one of the most remarkable paradoxes of our time.

Chapter 1

THE ECOSYSTEM

AMAZONIA DEFINED

The investigation of man's relationship to his environment can be conducted in two different ways: 1) a particular culture can be selected and the manner in which it is articulated with its habitat can be analyzed; or 2) a certain kind of environment can be chosen and the variability in cultural adaptation within its boundaries can be examined through time and space. Since no ecologically oriented field studies have been made as yet in the Amazonian lowlands, the second method will be adopted here.

Before attempting to analyze man's adaptation, it is necessary to define the limits of Amazonia. Selection of a satisfactory criterion requires recognition of the fact that not all aspects of the environment are equally significant for every kind of organism. On the contrary, during the course of its evolution, every species of plant and animal develops an intimate relationship with only a small segment of its total environment. Its niche may be defined in terms of elevation, chemical characteristics of the soil, nature of the food supply, maximum or minimum temperature, or innumerable other parameters. Although the limiting factors vary in kind and rigidity, no species is able to flourish equally well under all possible environmental conditions.

Identification of the environmental characteristics most relevant to human adaptation is complicated by man's ability to shield himself culturally from many conditions that are biologically adverse. One important sector in which culture cannot always completely counteract environmental deficiences, however, is in the quantity and quality of the subsistence resources. Plants and animals, even when domesticated, require specific combinations of heat, moisture, and nutrients; and although some kinds of deficiencies can be moderated culturally, others result from physical, chemical, and atmospheric conditions beyond human control. Since culture cannot attain more than a minimal level of complexity without a concentrated and productive food supply, differences in subsistence potential are the most significant aspect of the environment from the standpoint of human adaptation.

This being the case, the geographical extent of the Amazon watershed is obviously an unsuitable basis for demarcating Amazonia because the upper courses of the major tributaries drain regions that differ greatly in altitude, rainfall pattern, temperature, topography, and many other climatic and edaphic features that affect subsistence and especially agricultural potential. Nor can rainfall, temperature, or elevation be utilized singly, since their interaction influences plant growth in complicated ways. Wild plants are affected by the same environmental conditions as cultivated ones, however, so that the occurrence of a uniform type of primary vegetation provides an indication of general uniformity in those aspects of the environment most relevant to human adaptation.

In lowland South America, tropical rain forest prevails over an area of some 2,320,000 square miles, including most of the Amazon basin and extending northward over the Guianas to the mouth of the Orinoco (Fig. 1). Generally speaking, it is the dominant vegetation below 5,000 feet in elevation, where annual average temperature variation does not exceed 5°F., where rain falls on 130 or more days of the year, and where relative humidity normally exceeds 80 percent. Small enclaves of savanna appear where the soil is too porous to retain moisture during the dry season, but these are unimportant because they offer no special opportunities for human exploitation. In spite of its vast extent, the lowland tropical forest ecosystem constitutes a distinctive and remarkably homogeneous environment because of its long geological history, its uniform climate, and its equatorial location. Since these inorganic characteristics establish the boundaries within which biological and cultural adaptation take place, the story must begin with them.

THE INORGANIC ENVIRONMENT

Some 600 million years ago, in Precambrian times, the Guayana and Brazilian highlands (which occupy the northern and southern segments of Amazonia) were prominent mountains on a continent that has long since faded into oblivion. Millions of years of chemical and physical erosion have reduced the peaks to rounded hills and isolated plateaus and transformed the once rich soil into inert granite and white sand. During the Carboniferous the sea gradually withdrew, exposing low land drained by westward flowing rivers. Some 70 million years ago, the Andean mountain chain rose, causing the formation of an extensive freshwater lake, which occupied the central part of the Amazon basin throughout most of the Tertiary (Fig. 1). During both periods of inundation, sediments were deposited, ultimately accumulating to a depth of more than 6,500 feet. By the early Pleistocene, the eastern link between the Brazilian and Guayana shields had been sufficiently worn down to be ruptured. The outracing water sliced into the soft lake bottom with the result that even today the Amazon channel has a depth of more than 330 feet in several places below Manaus. During subsequent millennia, the drainage system of the Amazon basin gradually assumed its present form.

This geological history has produced a terrain of extraordinary flattness. The main river drops only 213 feet between the Peruvian border and the ocean, a distance of 1860 miles. Eleven of its tributaries flow for more than 1,000 miles uninterrupted by a single fall or rapid. In spite of this minimal gradient, the Amazon current maintains an average velocity of 1.55 miles per hour during the dry season; during high water the rate of flow more than doubles under the pressure of the tremendous volume of water deposited over the drainage area.

Its low elevation and equatorial location give Amazonia a remarkably uniform temperature. There is little seasonal variation in the length of the day or in intensity of solar radiation, and the mean temperature of the warmest month is only 5°F. above that of the coldest month. Greater variation occurs during a 24-hour period, particularly during the dry season when the mean high reaches about 90°F. and the mean low drops to about 70°F.

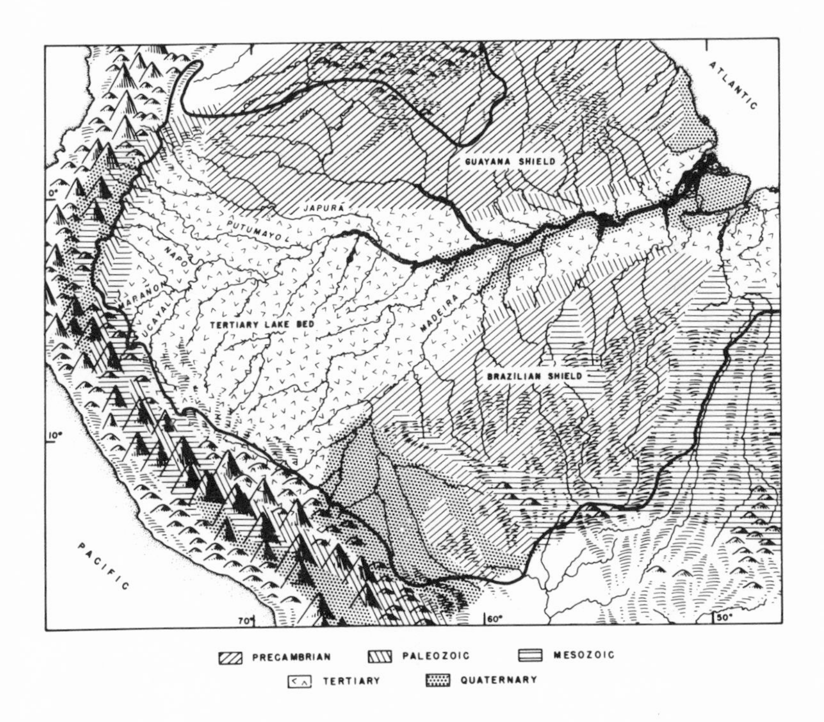

Fig. 1. Geological structure of the Amazon basin. The lowland tropical forest area is bounded by the heavy line. Its northern and southeastern portions are of Precambrian and Paleozoic origin, while the central and western part is composed of sediments deposited during the Tertiary. Soils of recent origin are restricted to the várzea, or floodplain, of the middle and lower Amazon and the headwaters of the Madeira in the south, where sediments carried down from the Andean highlands are annually deposited (after Gibbs, 1967, Fig. 2).

Warm temperature is coupled with heavy rainfall and high humidity. Precipitation exceeds 120 inches annually in the northwest and along the northeast coast, and most of the area receives in excess of 80 inches. No significant portion receives less than 60 inches (Fig. 6, p. 41). As the total annual precipitation declines, it tends to become increasingly seasonal, so that much of Amazonia experiences two or three months with little or no rain. Even during these "summer" months, however, relative humidity generally remains above 80 percent.

Description of tropical rainfall in terms of annual averages obscures two characteristics of major ecological significance: these are intensity and variability. About 20 percent of the rain comes in the form of cloudbursts, which have the capacity to release at least 0.04 inches of rain per minute for a minimum of five minutes. It has been estimated that the amount of water precipitated in such torrential downpours is 40 times greater than in temperate latitudes, constituting a formidable potential for leaching and erosion. The total annual rainfall and its monthly distribution may vary greatly from one year to the next. A wet year may receive four or five times more rain than a dry one, and even greater discrepancies have been reported. For example, the upper Essequibo area in southern Guyana received 1.63 inches in December 1954 and 12.03 inches in December 1955. While the wild vegetation appears not to suffer severely from such unpredictable fluctuations, they may have a disastrous impact on newly planted crops.

In spite of a rainfall three times higher than that of the central United States, the normal annual rise of the Amazon is only slightly over 32 feet, or half that attained by the crest of the Ohio as it flows past Cincinnati. This contradiction is explained by the equatorial course of the Amazon, which allocates the headwaters of northern and southern tributaries to different hemispheres. Rainfall intensity is regulated by the annual oscillation of air masses north and south of the equator following the seasonal variation in the position of the sun. Consequently, the height of the rainy season in the south falls between October and April, while the maximum in the north occurs between April and August. Since the southern rivers are falling by the time the northern ones begin to rise, their waters combine to produce a single crest on the Amazon (Fig. 2). Were this not the case, the annual influx

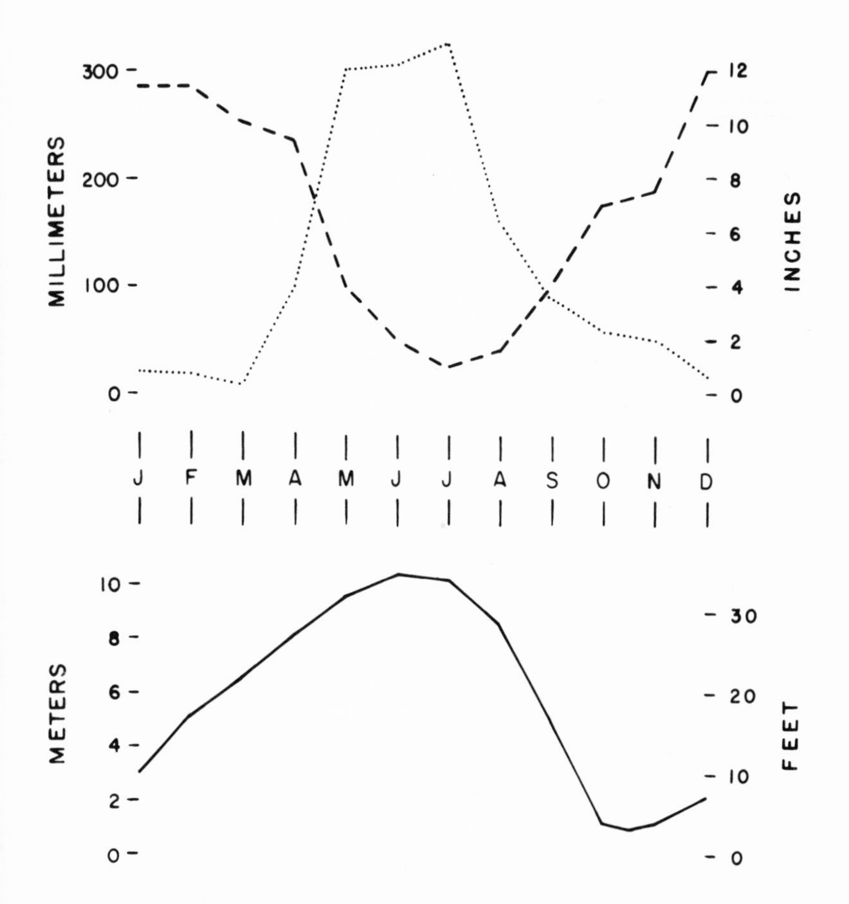

Fig. 2. Effect of the equatorial location of the Amazon on river rise and fall. In the northern hemisphere (dotted line), the rainy season extends from April through August; whereas in the southern hemisphere (dashed line), the wettest months are October through May. Since the southern tributaries rise as the northern ones begin to fall, a single crest is produced on the Amazon about the middle of June.

would be catastrophic and the floodplain ecosystem could not exist. The balance is delicate, and is often upset by fluctuations in onset or intensity of the rainy season in part of the catchment area. When the southern rains persist too long or those in the north begin early (as happens on the average every four years), land usually above high water level is inundated. Four times during the first half of this century, more severe dislocations of the "normal" rainfall pattern have occurred. Although the 1953 crest reached only ten feet above average, it had a disastrous impact on crops and cattle.

The combination of great geological antiquity, warm temperature, and heavy rainfall accounts for the remarkable infertility of Amazonian soil. In contrast to temperate regions, where physical weathering is the primary process of soil formation, chemical weathering predominates in the tropics. Warm rainwater percolating through the ground dissolves soluble minerals and carries them through the subsoil and ultimately into the rivers. The longer the process continues, the more the upper soil layers become impoverished, until only insoluble ingredients remain.

The low mineral content of most Amazonian rivers and streams attests to the advanced state of the leaching process throughout the lowlands. Two types of water predominate, both characterized by extraordinary purity and transparency but differing in color and other respects. The bleached sands of the Guayana and Brazilian shields give rise to rivers with dark brown water, which breaks in golden bubbles over rapids and falls. Such "black water" rivers have gradually sloping, poorly defined banks covered by periodically or permanently inundated forest. Litter falls from this vegetation into the riverbed where it decays, consuming oxygen and releasing free carbonic and humic acids in the process. The oxygen-deprived, acid, sterile aquatic environment of this flooded forest, or igapó, is one of the most remarkable and little studied ecological niches of Amazonia. From the standpoint of human exploitation, the black water rivers and the land they drain have such low subsistence potential that they are notorious throughout Amazonia as "starvation rivers." Among major rivers in this category are the Rio Negro and most of its right bank tributaries; and tributaries of the Tapajós, such as the Cururú and Arapiuns.

"Clear water" rivers resemble black water rivers in their low level of dissolved salts, absence of suspended silt particles, and tendency to acidity. Banks are high and stable, however, and the absence of oxygen-consuming organic matter in the water combined with greater transparency provides a more suitable environment for aquatic life, particularly where acidity is minimal, as it is in the Tapajós.

Least numerous but by far most important are the "white water" rivers, which gather a heavy soluble and suspended load of minerals and soil particles as they cascade down the Andean mountain slopes. Their muddy "white" water has low transparency and is neutral to slightly alkaline. The six major tributaries that tap the minerally rich Andes are the Japurá, Putumayo, Napo, Marañon (including the Huallaga), Ucayali, and Madeira (Fig. 1). Although their mountainous headwaters drain only 12 percent of the Amazon basin, they contribute 86 percent of the total dissolved salts and 82 percent of the suspended solids discharged by the Amazon. Some of this heavy burden of silt is deposited annually on the floodplain, or várzea, injecting a sliver of youthful soil into the ancient terrain.

No true lakes exist in Amazonia, and the numerous lake-like features that appear on or adjacent to the floodplain are river formations created by silting, channel alteration, or periodic inundation of low areas. "River lakes," identifiable by their attenuated proportions and tortuous shoreline, are produced by partial blockage of the mouths of clear and black water tributaries (Fig. 18). Water level is controlled by the Amazon, rising when the pressure of the Amazon crest inhibits flow from the constricted mouth and falling when the pressure is relieved. The high acidity and low nutrient content of the water in these lakes restricts the variety and density of aquatic fauna. Much smaller and far more numerous are the várzea lakes, formed in low places by the accumulation of rain and floodwater. Although the rainwater by itself is too pure and the white water of the Amazon is too turbid, mixture of the two creates an ideal environment for aquatic plants and for phytoplankton, which in turn support a fauna of remarkable density and diversity.

The Amazonian lowlands as they appear today are thus the product of millions of years of ecosystem evolution. Throughout much of the Pleistocene, erosion and leaching were favored by a combination of high average temperature and abundant precipitation. Ninety-eight percent of Amazonia consequently consists of terra firme, or upland, composed of geologically ancient soils drained by sterile black or clear water rivers. Only two percent is occupied by the várzea, which is annually rejuvenated by sediments carried down from the Andean highlands. The terra firme and the várzea represent two distinct habitats within Amazonia, and their importance for past and future utilization by man is inversely proportional to their relative sizes.

SUBSISTENCE POTENTIAL OF THE TERRA FIRME

Primary Determinants

Because the terra firme ecosystem is one of the most complex associations on the face of the earth, it would require several volumes to describe even the most important of the interactions and linkages among its climate, soil, flora, and fauna. Fortunately, such complete description is not necessary for an understanding of the principal problems of human adaptation. We need only to know what kinds of physical and chemical absolutes exist, because they represent the basic factors to which all plant and animal life, including man, must adjust if the species is to survive. Although many factors are involved, the most significant are the age of the soil and the characteristics of the climate, particularly temperature and rainfall.

In contrast to most soils of Europe and North America, which have developed since the beginning of the Pleistocene, the youngest terra firme soils date from the Tertiary. The Guayana and Brazilian shields are among the most ancient formations on earth. Millions of years of exposure to chemical weathering has leached out all soluble minerals, and the resulting "mature" soils consequently consist principally of sand and clay and are moderately to extremely acid. In terms of plant nutrients, the deficiencies are so severe that soils of similar composition would be barren in a temperate climate.

Another "absolute" is temperature, which affects several biological and chemical processes crucial for maintenance of soil fertility. For example, 77°F. is the critical temperature for the formation of humus, which performs a vital role. In sandy soils, it increases water-holding capacity and the ability to absorb plant nutrients; in clayey soils, it enhances porosity and permeability. In its absence, continuous agriculture is impossible. Humus accumulation takes place, however, only when soil temperature remains below 77°F. If temperature rises higher, bacterial activity increases to the extent that humus decomposition exceeds the rate of formation. Increased soil temperature also promotes the breakdown of humic matter into carbon dioxide, nitrogen, ammonia, and nitrate, large proportions of which are volatile and escape into the air.

The third "absolute" is rainfall, which acts both on the surface of the ground by erosion and on its internal composition by leaching. The erosive effects of water increase exponentially with increase in the rate of flow, so that if the flow is doubled, the scouring capacity quadruples, carrying capacity increases 32 times, and the size of the particles transported increases 64 times. Once erosion begins, a self-perpetuating and accelerating cycle is set in motion: removal of the absorbent surface layer increases runoff; increased runoff enhances carrying capacity; more particles in suspension aggravates the abrasive action; and increased abrasion removes more soil.

The combination of warm temperature and high rainfall affects the soil in other adverse ways. The low ratio of organic matter permitted by high temperatures enhances the solubility of silica and kaolin but promotes the retention of aluminum and oxides of iron. Iron is precipitated as lateritic concretions through a chemical reaction that also removes phosphorus, which is necessary for plant growth. Although laterization has the desirable effect of increasing resistance of the soil to erosion, it also lowers its ability to retain ammonia, lime, potash, and magnesia, all of which are important plant nutrients. Once laterite has formed, the process is irreversible.

In spite of these unfavorable characteristics of soil, temperature, and rainfall, the Amazonian lowlands support a magnificent forest vegetation. Trees average 50 percent taller than those in temperate woodlands, and the number of arboreal species is more than 20 times that in European forests. Tropical vegetation is often so diverse in composition that it is difficult for the expert to spot two trees of the same species in the same general area.

To the untrained eye, however, this diversity is masked by uniformity of appearance. The trunks are nearly all straight and slender, the base is often expanded into prominent buttresses, the bark is smooth, and the leaves are dark green, leathery, oval, and of similar size. Annual variation is also muted because of the small seasonal difference in weather.

The dense canopy of foliage formed wherever sunlight falls is composed not only of trees but also of quantities of climbing plants, many of which are epiphytic and trail their roots like streamers from the treetops. An impenetrable-appearing shield of vegetation is presented to the viewer, whether he looks down from an airplane or up from a dugout on the waterways. This dense and compact surface has led to the mistaken impression that a similar condition exists on the floor of the forest. On the contrary, in a primary forest the canopy of shade is so complete that undergrowth is kept at a minimum and a traveler frequently can pass through on foot with little clearing of the way.

Forests of this type flourish where the conditions of temperature and rainfall have all the detrimental features inherent in the climate of the lowland tropics. Because higher plants require a constant supply of soluble nutrients for normal growth and reproduction, and because these include relatively large amounts of nitrogen, phosphorus, potassium, calcium, magnesium, and sulphur, the maintenance of luxuriant vegetation implies success in overcoming (or at least minimizing) the negative effects of temperature and rainfall. This is exactly what the vegetation does, and the manner in which an equilibrium is achieved is instructive not only as an example of the intricate interaction between diverse components of the ecosystem, but as a basis for evaluating different cultural adaptations to the terra firme environment.

The way in which vegetation can mitigate or aggravate the impact of the climate on the soil can be illustrated by examining the major characteristics of four kinds of plant associations. The two extremes are represented by the primary forest and by total absence of plant cover. Intermediate positions are occupied by the two principal agricultural techniques: namely, the tropical variety involving mixed cropping among the trunks and branches that remain after the felled vegetation is burned; and the temperate variety involving clean clearing and planting a single crop.

Primary Forest

The remarkable success of the primary forest in offsetting the detrimental effects of the tropical climate is one of the most impressive accomplishments of natural selection. Almost every feature contributes to the conservation and recycling of nutrients and the preservation of an ecological balance.

The continuous canopy of evergreen foliage performs multiple functions, among them nutrient capture, nutrient storage, and protection of the soil from erosion and solar radiation. On the average, 25 percent of the daily precipitation is withheld by the leaves and the rest reaches the ground as a fine spray. The effectiveness of this protective covering is clearly demonstrated by the consequences of deforestation. Whereas an average annual rainfall of 85 inches removes less than half a ton of soil per acre on a forested 12 to 15 percent slope over a three-year period, 45 tons are lost from a denuded hillside only half as steep. The dense foliage also shields the soil from solar radiation, making it possible for a small amount of humus to accumulate and permitting completion of the nitrogen cycle, with the result that this critical element for plant growth is not dissipated into the air.

Perhaps the most spectacular accomplishment of the vegetation, however, is its ability to capture and store nutrients. Whereas in temperate climates nutrients can be accumulated in the soil until needed, under tropical conditions all elements not recovered immediately are vulnerable to leaching and permanent loss. The spectacular growth rate and vast bulk of tropical vegetation are adaptations for the rapid recapture and storage of nutrients. Recycling is facilitated by a rate of litter fall three to four times greater than that of New York woodlands, which return nutrients in proportions ranging from double in the case of phosphorus to over tenfold in the case of nitrogen. The foliage also plays an important role in nutrient capture from the air. Nearly 75 percent of the potassium, 40 percent of the magnesium, and 25 percent of the phosphorus available to growing plants are returned to the ground by rainwater dripping from the leaves.

Maximum utilization of the available nutrients is ensured by juxtaposition of plants with different requirements, and tropical forest vegetation is consequently characterized by a great proliferation of species but low concentration of individuals of the same species. Different kinds of plants not only have distinctive nutrient requirements, but they also vary in root character and penetration. The resultant dense root mat improves the structure of the soil during the life of the plants, and on their death makes available a significant amount of organic matter for processing into humus. Dispersed distribution of members of the same species also makes the plants less vulnerable to damage from predators and disease.

In summary, the primary forest of the Amazon basin, like primary forests in other lowland tropical regions, performs two major conservational functions: 1) it sets up and maintains a closed cycle of nutrients, so that the same ingredients are kept in continuous circulation and loss is reduced to a minimum; and 2) it mitigates the detrimental effects of the climate to the extent that soil impoverishment by erosion or leaching is either arrested or reduced to a very slow rate.

Clean Clearing

When all vegetation is removed, the soil is exposed to the full force of the climate. Falling rain compacts the surface of the ground, decreasing its penetrability. When absorption declines, runoff increases, enhancing erosion. Between showers, the sun raises the soil temperature to the point where the destruction of organic matter by bacteria outstrips the rate of formation, so that no humus can accumulate. The disappearance of humus lowers the water-holding capacity of the soil, and soluble minerals are rapidly carried down into the subsoil, where they are permanently out of reach of growing plants. Unshielded ultraviolet rays produce chemical changes in the soil, resulting in the conversion of nitrogen and carbon dioxide into gas which escapes into the air. Every degree that the temperature rises above 77°F. increases nitrogen loss by 15 to 20 pounds per acre per year. Loss of carbon dioxide and organic matter is not only detrimental in itself but leads to reversal of another process significant to the maintenance of fertility. When these ingredients are present, silica and kaolin are retained and oxides of iron and alumina are increased in solubility. When they are absent, silica and kaolin are lost, and iron oxide and alumina are precipitated as inert concretions of laterite, removing the available phosphorus in the process.

For as long as the surface of the soil remains exposed to sun and rain, heat and ultraviolet radiation perpetuate biological, physical, and chemical processes that inevitably reduce fertility. The addition of fertilizer, whether inorganic or organic, cannot raise the nitrogen content of the soil because of the rapid volatilization that occurs under exposure to sunlight. Removal of the forest vegetation thus initiates a series of events that either remove the soil or reduce it to sterility. The longer the interval between clearing and the onset of secondary growth, the greater will be the damage and the slower the rate of recovery.

Slash-and-Burn Agriculture

Throughout most of the lowland tropical forests, agriculture is conducted in temporary clearings, which are allowed to revert after a few years to secondary vegetation. This technique has been called slash-and-burn or shifting cultivation, terms that describe two principal features: 1) cutting and burning the vegetation prior to planting; and 2) shifting to a new clearing after two or three crops have been obtained. To observers from temperate Europe and North America, this method appears wasteful of labor and destructive of the forest, and proposals for increasing the productivity of tropical agriculture frequently specify replacement of the aboriginal system with permanent cultivation. Consequently, it is relevant to examine the relative success of the two systems in mitigating the harmful effects of the standard climate.

Although details differ, slash-and-burn agriculture in the Amazon basin has several constant features. These can be illustrated by a description of agricultural practices among the Mundurucú Indians of the Rio Tapajós. When selecting land for a new field, the Mundurucú prefer an area with a gently sloping and consequently well-drained surface, since too much water retention will cause root crops to rot in the ground. A clayey texture is recognized as superior to a sandy one. The size of the clearing is determined by the potential productivity of the soil and the number of people who will share the harvest. Fields tend to be circular unless the topography is incompatible with this form.

Clearing begins with removal of the shrubs and small trees, because these will be more difficult to cut after the large trees are felled and to give more "elbow room" for the latter operation. This preliminary cleaning takes about three days. The Mundurucú proceed to fell the large trees in a systematic manner. A particularly towering specimen at the higher edge of the slope is selected as a "keystone." Using it as the apex, a triangular zone is laid out and all trees in this zone are cut halfway through. Finally, the key tree is felled in such a way that it will carry down with it the adjacent partly cut trees. These in turn will bring down others until the entire area, which may have a width of more than 350 feet, is opened up. Any trees still standing, and those at the edges where they were out of reach, are cut individually. An average Mundurucú field is cleared in three days in this manner.

The vegetation is allowed to dry for about two months before burning. The fire is set on a day when there is a slight breeze to fan the flame, but not enough to sweep it so rapidly across the field that much unburned wood remains. Subsequently, incompletely consumed branches are collected into a pile for reburning.

Burning is timed to precede the first rains, and planting follows immediately. There is no cultivation or other disturbance of the soil. A hole is made with a digging stick, cuttings or seeds are inserted, and soil is moved over them with the foot. Usually about a dozen species of food plants are raised, with manioc and yam intermixed in the center of the field and other crops arranged in small plots around the edges. Weeding is done twice during the growing season. Harvesting of most plants takes place as the need arises, so that the entire crop is not removed at one time. Certain plants, especially manioc, may be replanted immediately after harvest to assure a permanent supply. Diminishing productivity leads to abandonment of a field after two or three years.

When this type of agricultural activity is analyzed in ecological terms, it is evident that it imitates the characteristics of the forest vegetation in several significent respects. The intermixture of crops with differing nutrient requirements, like the intermixture of tree species, reduces the competition for any particular element and makes maximum use of the range of nutrients available. The absence of large uniform stands also helps to protect against loss by disease, which spreads less easily when individuals of the same species are scattered and isolated. Weeding is a practice of mixed value: although it benefits the crop by eliminating competition for nutrients, it hastens soil deterioration by reducing shade and protection from erosion. Staggering the harvest, particularly if replanting is immediate, minimizes the time that the soil surface is exposed to the damaging effects of sunlight. Burning the slash returns some nutrients to the soil, which become available to the sprouting plants. When planting coincides with the first rains, nitrogen loss is minimized and the crops gain a headstart on weeds. The decomposition of trunks and large branches left lying on the ground returns additional nutrients at a slow rate so that they can be absorbed throughout the growing season. The presence of decomposing vegetation also diverts the attention of certain pests and diseases from the crop plants.

An imitation is never as good as the real thing, however, and in spite of its adaptive features, slash-and-burn cultivation is no match for the natural vegetation in offsetting the potentially destructive effects of the climate. The rapidly diminishing productivity of a typical terra firme field is dramatic evidence of this fact. In most cases, the second year's harvest declines only slightly, but the third year brings a marked reduction in yield, and the fourth year is generally so low as not to compensate for the effort of replanting (Table 1). The land is allowed to revert to forest, which undertakes the long process of restoring the conditions that existed prior to clearing.

Intensive Agriculture

The slash-and-burn method, despite its deficiencies, is far better adapted to the environment than the intensive agricultural techniques characteristic of temperate regions. Temperate agriculture involves clean clearing of fields, with all vegetation either removed from the site or gathered into one pile for burning. The stumps and large roots are eliminated along with trunks and branches, leaving the surface as bare as possible. Before planting, the field is plowed, serving both to bury weeds and to aerate the soil. A single crop is planted in even rows over a large area, and kept weeded until well established. At maturity, the crop is harvested within a few days, leaving the ground exposed to sun and rain. In the case of grains, the stems or stalks may be left and turned under in the spring plowing, after which the same crop may be replanted or another may be substituted.

When this system is used in the tropics, however, it has disastrous effects on the soil. Clean clearing immediately exposes the surface to the full intensity of the sun, accelerating the deterioration of both nutrient content and physical structure. Total removal of the preexisting vegetation prevents restoration

Table 1. Yields of major crops on terra firme and várzea soils[1]

Crop		Yield per 2.5 acres per year or harvest				
Terra firme	Várzea	1st	2nd	3rd	4th	5th
Maize		1,325–2,650 lbs.	1,650 lbs.	1,150 lbs.	Abandoned	Abandoned
	Maize	3,300 lbs.	3,300 lbs.	3,300 lbs.	3,300 lbs.	3,300 lbs.
Bitter manioc (12-month variety)		± 18 tons	±13 tons	± 10 tons	Abandoned	Abandoned
	Bitter manioc (6-month variety)	12–19 tons*	12–19 tons	12–19 tons	12–19 tons	12–19 tons
	Sweet manioc (6-month variety)	± 9 tons	± 9 tons	± 9 tons	± 9 tons	± 9 tons
Domesticated rice		2,650 lbs.	1,750 lbs.	1,325 lbs.	Abandoned	Abandoned
	Domesticated rice	6,600–9,900 lbs.*	6,600–9,900 lbs.	6,600–9,900 lbs.	6,600–9,900 lbs.	6,600–9,900 lbs.

*Variation in yield reflects differences in variety.
1. After Lima, 1956, pp. 108, 113, 154–157.

of any of the stored nutrients to the soil. Even hoeing destroys the favorable soil constitution, while deep plowing accelerates the breakdown of organic matter by increasing the available oxygen. By the time the crop germinates, many nutrients have already been leached away or volatilized and the plants must compete for those that remain. Any disease or predator can rapidly cover the entire field, damaging a major portion of the crop. In short, such methods not only destroy the soil beyond repair but also increase the risk of crop failure. There is no doubt that the slash-and-burn technique is better suited to tropical conditions.

The prevalence of slash-and-burn cultivation in the Amazonian lowlands thus represents an adaptation to the special requirements of the soil and climate. The fact that it is the only agricultural technique that can be practiced indefinitely without permanent damage to the land accounts for its widespread occurrence throughout the tropics. Its success in conserving the fertility of the soil carries with it a price, however, in the form of relatively low concentrations of population and permanency of residence. The manner in which these are reflected in the cultural adaptation will be examined in a later chapter. The point to be stressed here is that shifting cultivation is not a primitive or incipient farming method, but a specialized technique that has evolved in response to the specific climate and soil conditions of the tropical lowlands (Fig. 3).

Wild Food Resources

The varied natural vegetation includes hundreds of plants with edible roots, fruits, seeds, nuts, or berries, a large number of which are exploited by man. Many are consumed only as occasional "snacks" when they are accidentally encountered in the forest. Others are significant in the diet either throughout the year, as in the case of palmito or palm cabbage, or during the harvest season, as with the Brazil nut. Fruits generally ripen during the rainy season, before the newly planted crops are ready for harvest, and thus provide an important source of food at the time when it is most needed. Although potentially available in large quantities, wild plant foods are not concentrated because of the scattered and isolated distribution of species that is characteristic of the tropical forest vegetation. Exploitation of this resource consequently requires a relatively large expenditure of time and personnel as well as access to an extensive area of forest.

As a consequence of the scattered distribution of plant food, animals are also dispersed. With the exception of peccaries and monkeys, Amazonian animals tend to be solitary or to live in family groups rather than larger herds. The majority are small, and many are nocturnal. Among the principal game animals are the paca, which attains a length of 23 to 27 inches, the agouti and the armadillo (which are of similar size). The largest land mammal is the tapir, which frequents streams and lakes and reaches a length of 6.5 feet. Deer occur on or near savanna. Two species of

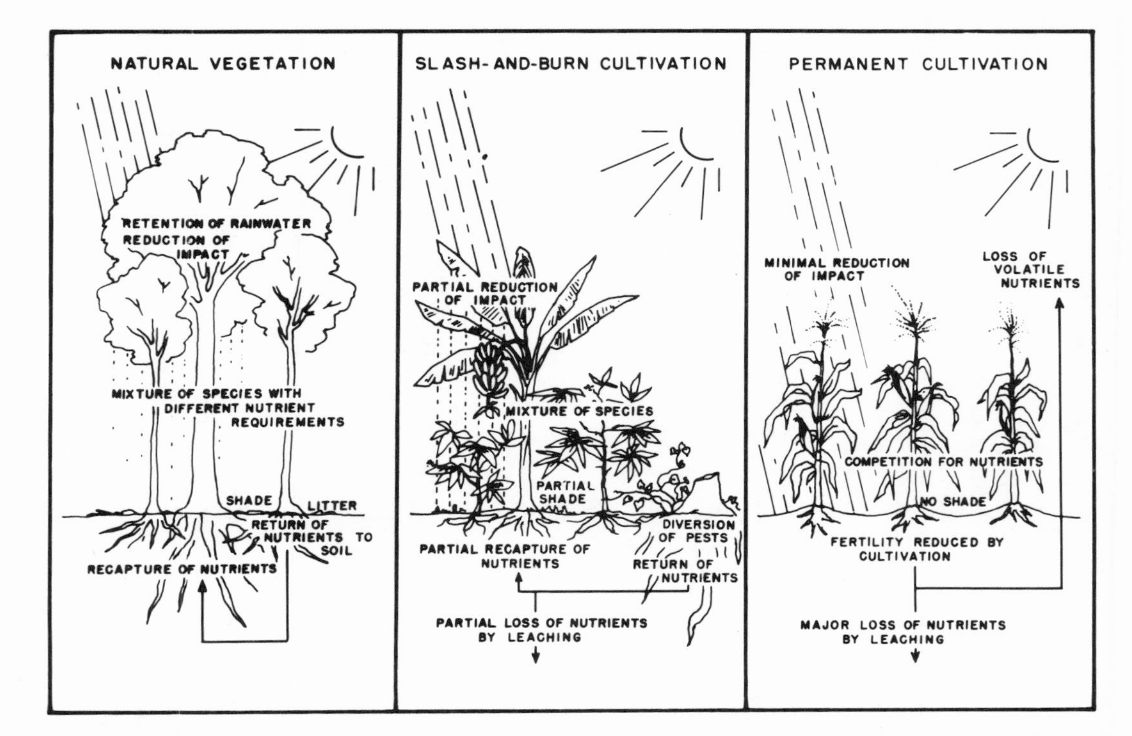

Fig. 3. Relative success of the natural vegetation, of slash-and-burn cultivation, and of permanent cultivation in minimizing the potentially detrimental effects of the tropical climate on the soil of the Amazon basin.

peccary grow to a length of about 40 inches and live in bands that sometimes number over 100 individuals. The most obtrusive forest animals are monkeys, which often can be heard conversing in the treetops. Land tortoises are abundant and one large variety reaches over two feet in length. Many species of birds are potentially edible; their sizes range from the large turkey-like curassow to small varieties, such as doves and toucans. Parrots and macaws are exploited for meat as well as for their brilliantly colored plumage. Other forest products include insects, grubs, and honey.

The most variable subsistence resource is fish. Black water rivers and small streams are low producers, as are many clear water rivers. Where nutrients are more abundant, however, fish may be numerous enough to constitute a significant part of the diet. Water turtles, caymans, and manatees are also available in the major rivers.

The low concentration of plant and animal foods available in the terra firme environment exerts a significant effect on the pattern of human settlement. A less obvious quality of tropical subsistence resources, however, is even more critical to successful adaptation: namely, the low nutritional content of most tropical plants.

Nutritional Considerations

The successful adaptation achieved by the vegetation to the harsh realities of impoverished fertility, high soil acidity, and other detrimental effects of the tropical climate makes the price of this adaptation easy to overlook. A moment's reflection, however, makes it obvious that leaves and roots synthesized mainly from water and sunlight must be deficient in vitamin and mineral content. While this is of little importance to the vegetation cycle, it has critical implications for faunal development. Plant-eating animals are directly affected, and carnivores (including man) indirectly.

In temperate climates, with low average temperature and rainfall, soils tend to have relatively high inorganic fertility. The availability of mineral nutrients, especially calcium, favors the development of vegetation with minimal bulk and maximum protein value (Fig. 4). Protein is particularly concentrated in the seeds, by which reproduction takes place. Such plants provide a nutritious diet to animals, some of which specialize on leaves (including grass), others on fruits or seeds. Temperate plants also often grow in large concentrations, furnishing an abundant harvest within a small area. The existence of plentiful and wholesome plant food has favored the evolution of herbivores with large body size (bears, moose, deer) and gregarious habits (bison, caribou, horse).

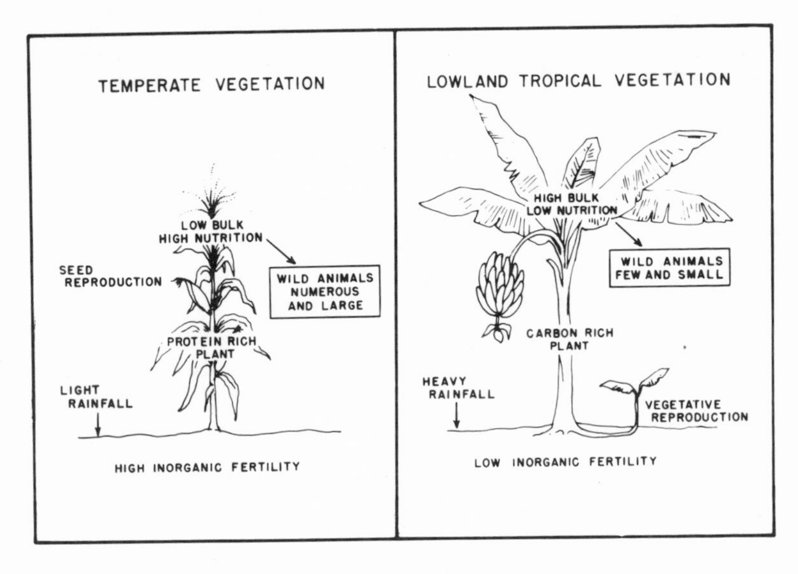

Fig. 4. Effect of climate, soil, and rainfall on plant reproduction and nutritional content.

In the tropical lowlands, however, conditions of high average temperature and rainfall are coupled with low inorganic fertility to produce vegetation with large bulk and limited protein content. In the absence of sufficient protein for seed production, many tropical species have evolved vegetative methods of propagation (Fig. 4). Leaves and grass developed under these conditions are luxuriant in appearance, but their limited food value is reflected in a fauna that is sparsely distributed and small in body size in comparison to that of temperate regions. Browsing animals are rare and most herbivores subsist on fruits, nuts, tubers, and water plants, which have higher nutritional value relative to bulk than do leaves. Even so, the indigenous fauna has had to adapt to a lower level of protein intake than temperate animals are able to tolerate.

The fact that human populations indigenous to the lowland tropics have also become adapted to low levels of protein intake is evident in the fact that they show no adverse effects although their consumption is below temperate standards. Furthermore, tropical peoples have a capacity not found in temperate inhabitants to store protein in their bodies for several weeks. This physiological adaptation accounts for the widespread practice among tropical hunters of consuming vast quantities of meat at one time. Similar physiological adaptation has been achieved with regard to other scarce elements, among them calcium, certain vitamins, and salt. As a consequence, deficiency diseases are rare among tropical peoples whose subsistence has not been degraded by recent modifications, although they quickly develop among immigrants from the temperate zone who subsist on the same diet.

Although tropical peoples have adapted physiologically to a comparatively low nutritional level, they still require minimal amounts of basic elements. Tropical cultigens are particularly deficient in protein, and even maize and rice contain fewer nutrients when grown under tropical conditions. Certain wild plants, however, have achieved abnormally high nutrient concentrations in their fruits and seeds. An outstanding example is the Brazil nut, which contains 50 percent more protein per 100 grams than maize. A balanced diet therefore cannot be obtained without combining cultivated staples with wild fruits and nuts, game, and fish. Over the millennia, each aboriginal group succeeded in developing a seasonal cycle that combines hunting, fishing, gathering, and agricultural activities of different kinds and relative intensities, but which in every case assures the availability of all essential nutrients indefinitely without endangering the equilibrium of the ecosystem. Tendencies toward overexploitation of any particular resource are controlled by a variety of cultural practices, some of which seem at first inspection to be irrelevant to adaptation. More careful examination indicates that they constitute fascinating and effective solutions to the problem of long-term survival of the group.

SUBSISTENCE POTENTIAL OF THE VARZEA

Primary Determinants

Although the várzea occupies the heart of the Amazon basin, where the tropical climate reaches its maximum expression, it differs in two significant ways from the terra firme. First, its soil is annually rejuvenated by a layer of fertile silt of Andean origin; second, its annual cycle is determined by rise and fall of the river rather than the seasonal distribution of local rainfall.

The only rivers that transport significant amounts of fertile sediment are the Marañon and Ucayali in highland Peru, the Madeira originating in Bolivia, and the Napo, Putumayo, and Japurá flowing out of Ecuador and Colombia. For reasons that are largely topographical, most of the silt is carried to the Amazon floodplain before it is released. Consequently, although narrow strips occur along the lower portions of the Madeira and Purús, the várzea is predominantly a formation of the middle and lower Amazon (see, Figs. 1, 17), where it occupies an area of some 24,000 square miles. As the river enlarges, the várzea also expands. Above the Rio Negro, its width is typically less than 16 miles. Between the Rio Negro and the first major islands above the mouth, width increases to 30 miles. At the "delta" it reaches a maximum of about 125 miles.

In spite of its comparatively small area, the várzea is a complex and heterogeneous environment, mainly because the silt-bearing waters distribute themselves unevenly over the floodplain. The factor principally responsible for this inequality is the priority of the local rainy season climax over the annual river crest. Along the middle and lower Amazon, the rains begin in November or December and reach their peak between March and April. During this period, surface runoff, drainage water, and precipitation contribute water of relatively high purity to the depleted várzea channels, ponds, and lakes. Since the Amazon does not attain its crest at Manaus until more than two months after the local rainy season has begun to abate, its spreading waters reach many lakes only after they have already been completely or partly filled. Interplay between these two different kinds of water creates a hodgepodge of clear, black, white, and mixed water lakes, ponds, and channels that provides an extraordinary variety of conditions for aquatic plant and animal life.

The silt-laden flood waters not only distribute themselves irregularly over the várzea, they also deposit their sediment unevenly. The largest particles settle on the river bank, creating a gradually rising ridge averaging about 500 feet wide (Fig. 5). This levee or high várzea is better drained and more briefly inundated than

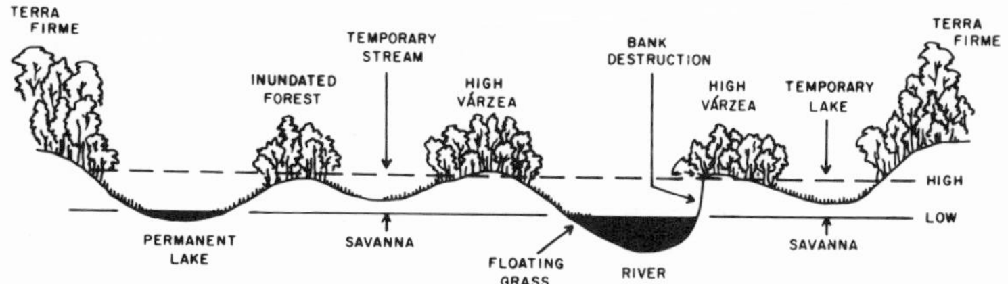

Fig. 5. Idealized cross-section of the várzea showing the types of niches created by alternation between high and low water. The vertical dimension is exaggerated.

the hinterland or low várzea, which may remain submerged or waterlogged throughout the year, and it is generally well suited to agriculture. Its fertility is far from uniform, however, because sediment deposition is affected by the contour of the channel and the speed of the current, both of which are constantly changing as the river cuts away one bank and builds up the opposite one. Local differences in turbulence also interfere with even sedimentation, with the result that soil composition may vary greatly within a small area. Furthermore, when levees have become sufficiently elevated to remain above flood level in normal years, the soil becomes susceptible to leaching and consequent rapid decline in fertility under intensive agricultural use. Where circumstances are favorable, on the other hand, 2.5 acres of várzea receives annually about 9 tons of sediment containing ample amounts of sodium nitrate, calcium carbonate, magnesium sulfate, superphosphate, potassium chlorate, and other essential plant nutrients.

The regime of the Amazon is ideal for agricultural purposes. In contrast to the Nile, which rises rapidly and falls slowly, the Amazon takes eight months to reach its maximum but only four months to return to minimum level (Fig. 2). During April it rises about two inches per day, diminishing in speed during the last two months before climax. By mid-June, it begins to drop, and during September it falls five inches per day. As a consequence, a farmer generally has time to harvest his crop before inundation of the field, and the land is released by the river swiftly enough for good drainage before replanting. Normal flooding inundates the high várzea to a depth of only a few inches, although crests six or seven feet above normal occur at irregular intervals. At Manaus, the difference between average high and low water levels is about 32 feet.

In short, compared with an equal area of terra firme, the várzea is a diversified and variable place. From the standpoint of human utilization, it offers both natural food resources and possibilities for agricultural exploitation that are superior to those of the surrounding terra firme.

Natural Vegetation

The natural vegetation of the high várzea is forest, but its composition differs from that of the terra firme. Only species tolerant of periodic inundation can grow, and palms are consequently abundant. The low várzea, which is submerged annually, supports either forest or grass, depending on such variable local characteristics as soil, slope, current, depth of water, and duration of immersion. Generally speaking, lakes and ponds are fringed by savanna, which creeps down the banks as the water retreats and rides back in the form of floating rafts as the water rises. At certain times of the year, dense mats of floating grass completely camouflage the surface of lakes and channels. The roots of these plants provide an abundant food supply for aquatic fauna. Similar floating meadows form in stretches of calm water along the shores of clear water rivers and lakes, but do not develop in black water.

Permanent bodies of quiet water also support a variety of aquatic vegetation. The most striking component is the *Victoria regia* lily, whose cartwheel-sized leaves with upturned edges float on the surface like huge platters, interspersed with massive, succulent white flowers.

Agricultural Practices

The problem of clearing fields on the várzea is essentially the same as on the terra firme. Recent experiments along the Rio Guamá near Belém have provided detailed information on the number of man-days required for each step of the operation on an average hectare (2.47 acres) of mixed high and low várzea. An area this size contains about 512 trees, of which 80 percent are less than 20 inches in diameter. Thirteen percent have diameters between 20 to 40 inches, and the remainder exceed 40 inches in trunk diameter.

Clearing begins with removal of vines and saplings, a task that consumes six days of labor. Two days are required for girdling the açacú trees, which have a caustic sap that is potentially injurious. After a delay to allow the slash to dry, tree removal begins. The trunks are notched to make them fall in such a way that coverage of the ground will be relatively uniform. Felling the largest tree initiates a "domino" effect when the notched trees in its path give way under pressure. Removal of large branches from the prone trunks is the final step. Notching and trimming require 20 days of labor.

Burning takes place about a month and a half later at noon on a day with a clear sky and a good breeze, following four days without rain. Twigs and leaves piled along one edge are ignited first, to give the fire a good start. If the burning is done skillfully, only the largest trunks and branches remain unconsumed; cutting and stacking these requires another 20 days of labor. Stump removal is a tedious process requiring the investment of 377 man-days, a labor cost that is not compensated by an appreciable increase in the amount of productive land.

The crucial factor in successful farming of the várzea is timing the planting. If cultivation takes place when the ground is too wet or too dry, it has detrimental effects on both soil structure and plant germination. Certain crops (manioc among them) prefer well drained soil and consequently do best on high várzea. Maximum land use is achieved if the planting continues for several weeks, during which the field is progressively extended down the slope behind the retreating water. The lower limit for manioc occurs at the point where less than eight months intervene between exposure of the land and its subsequent inundation. Since maize matures in 120 days, it can be grown at lower elevations. When planted on high várzea, however, it can yield two harvests per year. Rice, a cultigen of European introduction, will also produce two harvests annually.

Because fertility of the soil is annually renewed by sedimentation, várzea fields do not decline in productivity with continuous use. Furthermore, yields are often two or three times greater than even the initial harvest from a comparable area of terra firme (Table 1); when dual cropping is practiced, the discrepancy is even more marked. In these two important respects (namely, permanency of exploitation and maintenance of high yield), várzea agricultural potential compares favorably with that of temperate regions.

Wild Food Resources

Wild food resources are also more concentrated and productive on the várzea than on the terra firme. As the water recedes, several species of aquatic grasses proliferate along the margins of lakes and streams, providing an abundant crop of seeds that attracts vast numbers of birds. Wild rice grows in dense stands.

It is in the water, however, that the real wealth of the várzea lies. Over 100 years ago, the famous naturalist Louis Agassez remarked on the tremendous diversity of the fish population:

> The Amazon nourishes about twice as many species of fish as the Mediterranean, and a more considerable number than the Atlantic Ocean from one pole to another. All the rivers of Europe combined, from the Tagus to the Volga, do not feed more than 150 species of fresh water fish, and yet in one little lake in the neighborhood of Manáos, called Lago Hyanuary, which has an area of 500 square yards, we have discovered more than 1200 distinct species, the greater part of which have not yet been observed elsewhere (Gonçalves, 1904, pp. 38–39).

The fish of the Amazon are remarkable not only for their variety, but also for the large size of many species. The piraíba, a catfish, attains a length of nearly 10 feet and weighs over 300 pounds. Several other species of catfish exceed 3 feet in length. The largest scaled fish is the pirarucú, which frequently measures more than 6 feet and weighs over 170 pounds. The pescada and the tucunaré, two of the most palatable species, are about 24 inches long.

Although fish are obtainable throughout the year, they are particularly easy to capture in quantity when the water level is low. The pirarucú, being large, easily caught, and palatable either fresh or dried, has been a primary target of commercial exploitation. In 1953 alone, more than five million pounds reached the market, while additional thousands of pounds were consumed by local residents. Commercial fishing at the confluence between the Rio Negro and the Amazon nets a daily catch of about 1,000 pounds between dawn and noon.

Water turtles are another extremely abundant food resource. The largest species is the tartaruga, which attains a length of three feet and an average weight of 50 to 80 pounds. Females lay between 100 and 150 eggs in a shallow pit on a sandy beach when the water level is falling. Both the turtles and their eggs are consumed in quantity by the local population and have been exploited commercially since the colonial period. The number of eggs destroyed annually for production of oil is estimated between 33 million and 72 million, representing the output of 330,000 to 480,000 females. Innumerable additional eggs are destroyed by birds, caymans, and other predators, which also feed on newly hatched turtles. Thousands of adult turtles are captured annually when they congregate on the beaches to lay their eggs, and a good hunter can obtain 10 to 15 a day during other times of the year. Such figures lend credence to the Indian tradition that in aboriginal times the middle Amazon seethed with tartarugas like an ant hill with ants. A smaller species, the tracajá, measures 15 to 20 inches in length and lays 25 to 30 eggs. It is said to be superior in flavor to the tartaruga, but does not thrive in captivity. The matamata, which has the dubious distinction of being the most hideous Amazonian reptile, attains a length of two feet. Because it prefers muddy places, the matamata is difficult to capture in large numbers.

The Amazonian waters are also inhabited by several species of aquatic mammals. The most important is the manatee, which attains a length of ten feet and a weight of 2600 pounds. It is prized not only for its meat but also as a producer of oil. The carnivorous freshwater dolphin rarely exceeds 6.5 feet in length or weighs over 330 pounds. Although consumed in aboriginal times, it has more recently become a target of superstition and consequently is deemed inedible. Manatee meat and oil, on the other hand, were important commercial products during the colonial period; the Dutch alone are reputed to have exported more than 20 shiploads annually.

The abundance of small fish and wild grass seed in the várzea lakes during the dry season once attracted ducks, herons, storks, and other kinds of birds in astonishing profusion. Travelers along the middle Amazon near the beginning of the nineteenth century reported sandbars covered with nesting birds so extensive that a canoe required half an hour to pass from one end to the other. As recently as the 1950's "colossal bands" frequented the lakes near Manaus and flew over the city in thick clouds.

The birds and their eggs in turn provided abundant fare to the alligator-like caymans, which awaited them at the water's edge with half-open jaws. The largest species attains a length of 13 to

16 feet and several smaller species reach 6.5 feet. Both the reptiles and their eggs are eaten.

A few decades ago, the appearance of the várzea in the dry season was graphically described as follows:

The mornings at this time of the year are incomparable in their lovely luminosity, refreshing coolness, and Dionysian animation. This is the season of abundance and plenty. In the lakes, inner rivers, and innumerable threads and cords of that endless texture of waters unique on this earth, occur in superabundance the pirarucú, the tucunaré with its brilliant rainbow colors. . . . the pescada, the tambaqui, the camorim, the aruanã, and the manatee, which makes a delicious conserve; in short, the whole numerous fish family, which it is both tedious and difficult to enumerate. It is also the season of those tasty morsels, the tartarugas, the tracajás, the mussuans and other kinds of turtles. And to climax this unequaled pantheistic scene, a multitude of gay fluttering birds ranges over the expanding margins of the interminable marshes in whirling flocks, curdling the subtle and imponderable turquoise blue waters.

The whole panorama is a close facsimile of that other Paradise of biblical times (Mendes, 1938, p. 34, trans.).

Nutritional Considerations

Although quantitative comparisons are not available, there is no doubt that the nutritional value of várzea plants and animals is higher than that of terra firme flora and fauna. Furthermore, the várzea is suitable for the cultivation of maize, which is a more concentrated source of minerals and vitamins than manioc and sweet potatoes. Animal protein is also available in high density and inexhaustible supply under aboriginal methods of exploitation. All of these factors operate to make the subsistence potential of the várzea far superior to that of the terra firme for human utilization.

The várzea is not an unmitigated paradise, however. At unpredictable intervals, the river rises 6.5 feet or more above normal, and the consequent sudden depletion of the food supply would have traumatic effects on a human population dependent on optimum subsistence conditions. Adaptation thus favored stabilition at a level compatible with a lower carrying capacity. This situation placed an upper limit on population size, although the ceiling was much higher than that prevailing on the terra firme.

MAN'S ARRIVAL

The Amazonian tropical forest ecosystem had reached maturity long before man made his appearance. While it now seems probable that he entered the New World via the Bering Strait at least 30,000 years ago, the oldest traces of his presence yet encountered in southern South America are less than half that age, and those from the lowland tropical forest are still more recent. The failure to discover ancient sites in Amazonia has a number of obvious explanations. First, the absence of suitable stone makes it probable that tools and weapons were made of wood and other perishable materials, which have little chance of survival. Second, only settlements that remained for many years in one place would produce a noticeable modification of the soil to mark their location. Third, the continual oscillation of the river channel, coupled with the annual alluviation, makes it unlikely that land suitable for settlement a few thousand years ago would still be accessible even if it remained intact. Fourth, the dense vegetation masks the ground surface, concealing any artifacts that might lie upon it. It is no accident that most existing evidence of early man in the New World has come either from areas that are now semidesert, where even minor surface indications are easily visible, or from rock shelters where refuse is concentrated and protected from natural disturbance.

The earliest dated human remains so far discovered in South America are from the coast of Venezuela, where stone artifacts in association with extinct fauna have a carbon-14 age of about 14,000 years. Rock shelters in the highlands of Colombia and Peru have produced cultural remains dating at about 12,000 years ago and man reached the coast of Chile only a few centuries later. The hunters and gatherers who made and used the stone and bone tools found at these campsites probably lived in bands composed of related families and wandered from place to place as the local subsistence resources were depleted or seasonal plant foods became available.

People dependent on wild foods would certainly have found the várzea an ideal habitat. The abundance of fish, turtles, manatees, and water birds, along with wild rice, cashew, cacao, and many varieties of palms with edible fruits would have offered a bountiful subsistence even in the absence of cultivated plants.

Unfortunately, it is not yet known how long the várzea has existed in something approximating its present condition, and until more specific information becomes available it is useless to speculate on the probable time of man's arrival in the lowlands.

The earliest carbon-14 date yet obtained from Amazonia is 980 B.C. (SI-385). It comes from a pottery-bearing site on the Island of Marajó, indicating that settled village life had been established on the lower Amazon by this time. The refuse accumulations of this Ananatuba complex are small enough to represent a single circular communal house, and occur in patches of forest adjacent to a small stream. The archeological deposit is thicker than that of later sites on the terra firme and contains an unusually large amount of broken pottery. Both of these features indicate that the village remained in one place for a considerable period of time. The pottery, which is the only surviving element of the material culture, consists of small rounded bowls and large globular jars. While decoration usually consists of brushing or scraping marks covering the exterior surface, a few vessels were ornamented with broad incised lines, sometimes defining zones filled with fine cross-hatching. The fact that this type of zoned incision was popular several centuries earlier in the Andean area makes it probable that the Ananatuba Phase population derived its knowledge of pottery making from the west.

An attempt to reconstruct the subsequent aboriginal history of the Amazon basin is frustrated by the scattered and vague nature of the archeological information. Intensive investigations have been conducted in relatively few places, most of them toward the margins of the area. Fortunately, more is known of the archeology of Peru, Ecuador, Colombia, and Venezuela, and these sequences along with recent finds in Mesoamerica help to place Amazonia in the framework of New World cultural development, particularly with reference to the origin of the staple food plants.

After decades of ill-founded speculation, during which regions as different environmentally and as widely separated geographically as Paraguay and Mexico were proposed as the place of origin, it has finally been established that the process of maize domestication began in Mexico before 5000 B.C. During subsequent millennia, maize cultivation spread northward and southward, reaching the Peruvian coast between 1400 and 1200 B.C. While there is no direct evidence of its introduction into the

Amazonian lowlands by this date, the fact that maize is associated in Mesoamerica and the Andean area with zoned incised decoration of the type employed by the Ananatuba Phase potters suggests that these two elements may have entered the tropical forest together at around 1000 B.C.

The time and place of domestication of the staple root crops of the Amazonian lowlands—manioc and sweet potatoes—is still unknown. Eastern Brazil and the llanos of the Orinoco are favored localities, but the supporting evidence is inconclusive. Wild species of the genus *Manihot* occur in frost-free habitats all the way from northern Mexico to northern Argentina. Manioc was added to the roster of cultigens in Mexico and Peru around 1000 B.C., and indirect evidence in the form of large pottery griddles resembling those used today for making cassava bread implies that bitter manioc was a staple food on the Caribbean coast of Colombia a few centuries earlier.

As a potential place of manioc domestication, the floodplains of the lower Magdalena, Cauca, and Sinú rivers on the Colombian coast have several points in their favor. Ecologically, they resemble the várzea of the Amazon. The streams, lagoons, and swamps abound with the same kinds of aquatic animals, including turtles, caymans, and fish. Unlike Amazonia, however, this region was in early communication with Mesoamerica, where the transition from wild to domesticated food plants was under way, and from which the idea of domestication could have been introduced if it did not develop locally. Although existing carbon-14 dates are too few to be conclusive, they suggest that bitter manioc utilization is oldest here, and later spread into the Amazonian lowlands. The relative recency of griddles in the Amazon basin is also consistent with this inference.

The uncertainties surrounding the origin and diffusion of tropical cultigens make it obvious that any reconstruction of the evolution of aboriginal cultural adaptation to the tropical forest must be regarded as mostly fiction. What we know of cultural development in other parts of the world, however, as well as the obviously tight integration that had been achieved between the aboriginal culture and its environment by the time of earliest recorded observation, permits us to postulate a long period of adaptive evolution. While certain cultural elements, among them pottery making and maize (and possibly also manioc) cultivation, were introduced from outside the area, most features of material culture, sociopolitical organization, and religion were certainly evolved locally or reshaped in response to selective pressures of environmental and cultural origin. Throughout the millennia preceding European contact, the extensive network of navigable rivers and the relative uniformity of the ecological setting facilitated movement of both people and cultural traits from one margin of the basin to the other. The patchwork occurrence of Arawak, Carib, and Tupí languages is one consequence; the extensive distribution of several ceramic traditions is another. With the passage of time, the process of adaptation produced a unique cultural configuration, which not only permitted the satisfaction of basic human needs with a minimum of effort, but was in harmony with the rest of the ecosystem. An examination of five aboriginal groups characteristic of the terra firme, followed by a reconstruction of two inhabiting the várzea at the time of European contact, will illustrate some of the methods by which this equilibrium adaptation was achieved and maintained.

Chapter 2

ABORIGINAL ADAPTATION TO THE TERRA FIRME

While the terra firme of Amazonia is an environment characterized by high rainfall, warm temperature, and impoverished soils, none of these features is uniform throughout the area. Regional differences in geological history, elevation, topography, rainfall, and vegetation affect the character of the soil. Precipitation varies in intensity, frequency, and monthly distribution not only from one part of Amazonia to another, but within the same region from one year to the next. Not all plants and animals are equally successful in exploiting these differing conditions, and natural selection over many millennia has favored the dominance of those species best able to adapt. Consequently, although flora and fauna are similar throughout the terra firme, significant regional differences are likely to exist in the accessibility and abundance of the component species.

If adaptation is a major determinant of culture, then the aboriginal cultures should reflect this environmental situation in two principal ways. First, there should be a general cultural pattern throughout the tropical forest area, which is a response to the generalized characteristics of climate and soil that define the region as a whole. Second, there should be local variations in subsistence emphasis, settlement size, and other cultural features that are correlated with local differences in the availability and abundance of subsistence resources.

These hypotheses can be tested by analyzing and comparing the cultures of several aboriginal Amazonian groups that inhabit subregions with slightly different rainfall, topography, and subsistence resources. Before adaptation can be considered to be the principal cause of cultural resemblances, however, two other possible explanations must be eliminated. Since similarities often result from diffusion, it is preferable to select groups too widely separated geographically for significant recent communication between them to have occurred. Similarities resulting from common ancestry are more difficult to recognize, but, other things being equal, groups that speak unrelated languages are less likely to share a common origin than those with linguistic ties.

The five groups selected as examples of terra firme adaptation appear to fulfill the requirements of independent origin and mutual isolation. The Kayapó and Camayurá live near the southeastern margin of the tropical forest (Fig. 6), a region that experiences a well-defined dry season for about three months each year. The Ge-speaking Kayapó are forest dwellers, subsisting mainly by hunting and gathering, while the Camayurá are river bank farmers and fishermen who speak a language belonging to the Tupi-Guaraní family. The Sirionó, who inhabit the periodically inundated lowlands of eastern Bolivia, also speak a Tupian language, but differ markedly from the Camayurá in their way of life. The western margin of the terra firme, where rainfall is heavy throughout the year, is the habitat of the Jívaro, whose language has no close relatives and who subsist principally on game and cultivated plants. The Cariban-speaking Waiwai who live in northeastern Amazonia, where rainfall also tends to occur the year around, follow a subsistence pattern similar to that of the Jívaro but emphasize different staple crops.

An attempt to analyze the aboriginal adaptation attained by these five cultures is complicated by several factors. Among them is the extent to which pre-Columbian population size and concentration have been affected by European contact. All of the sample societies have suffered decimation, and this recent population decline must be kept in mind in assessing the role of certain cultural practices. Behavior that appears detrimental to the survival of the group under present circumstances could have been adaptive during the pre-European period when disease and other biological controls on population size and density were less severe.

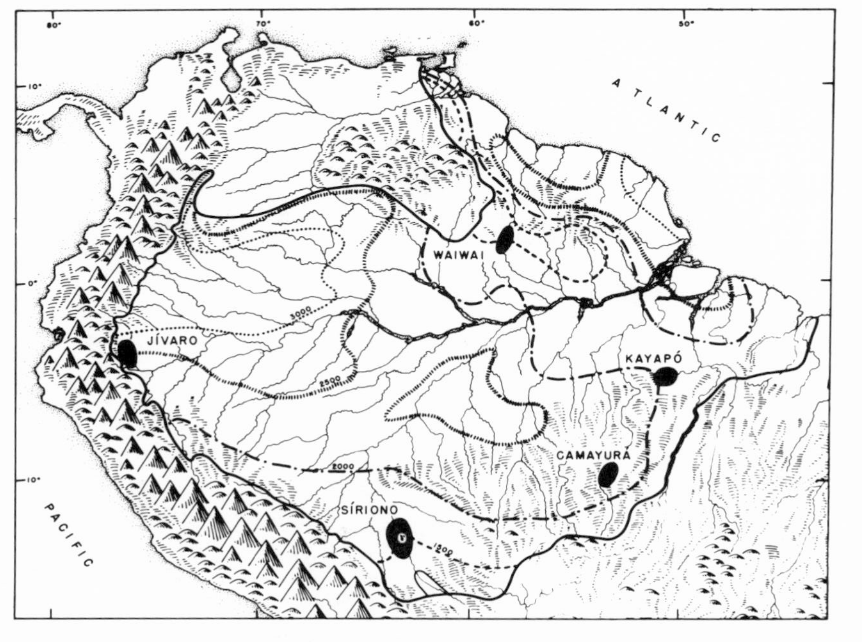

Fig. 6. Location of the five tribes selected to illustrate cultural adaption to the terra firme environment, in relation to the pattern of rainfall. Average annual precipitation exceeds 120 inches (3000 mm.) at the eastern and western margins of the area and 80 inches (2000 mm.) elsewhere, except in a narrow band slanting across the lower Amazon and skirting the southern boundary of the basin. The heavy line defines the limits of Amazonia.

In addition to population decline, most terra firme groups have experienced some degree of acculturation as a result of contacts with representatives of European civilization. Fortunately, relatively complete ethnographic descriptions were obtained for the groups in our sample while the effects of this contact were still superficial. European goods, such as knives, guns, glass beads, and mirrors, had been added to the material culture inventory without significantly affecting aboriginal arts and crafts. A few new food plants had been adopted, particularly bananas and sugar cane, but these did not replace indigenous staples. The most marked alteration is a reduction in the intensity of warfare, which has occurred partly because of population decline and consequent general cultural deterioration and partly because of policies of suppression by the national governments.

By sorting out such post-European innovations, it is possible to assemble a general description of the aboriginal way of life of the Camayurá, Jívaro, Kayapó, Sirionó, and Waiwai that includes the principal details of their subsistence round, settlement pattern, material culture, social organization, life cycle, religious practices, and relations with neighboring groups. Analysis and comparison of these characteristics should reveal differences in pattern that are attributable to environmental variables. In order to facilitate this comparison, descriptions follow a standardized outline and digress as little as possible from the essential facts. While elimination of many interesting "human" details may reduce the readability of these sketches, it also permits attention to focus on those cultural elements most relevant to the present inquiry.

The description of a culture, like the description of an animal species, provides those details that characterize a particular group and distinguish it from other similar groups. Since flexibility is a requirement for adaptation to continually changing conditions, all species and all cultures exhibit more or less variation through time and space. Type descriptions distill this variation into a series of generalized statements or rules. Like all rules, some are inviolate while others tend to be disobeyed. Among the latter are rules regarding the sexual division of labor, since the illness, absence, or death of a spouse frequently puts a man or woman in a position where it becomes desirable to perform a task normally assigned to the opposite sex. Rules of marriage are more binding, but even here occasional lapses may be permitted if a person would otherwise be unable to secure a mate and thus to become a full member of the society. Every culture has certain areas, however, where deviations are punishable by death, implying that flexibility would threaten the integrity of the system and consequently cannot be tolerated. In Amazonia, these tend to be concentrated in the category of ritual practices, which women and uninitiated boys are strictly prohibited from observing. Adultery may also carry a death penalty.

Cultural descriptions share another defect with biological descriptions: both must resort to arbitrary categories to organize data. Animals are described in terms of reproduction, locomotion, respiration, metabolism, and other component systems; while cultures are fragmented into settlement pattern, social organization, religion, and so on. In both cases, it is important to keep in mind that such descriptive categories are not independent systems. Not only do they supplement and reinforce each other, but often the same structures are involved in more than one system. Both animals and cultures are highly integrated organisms in which some parts are dominant over others but in which all parts must be compatible if the organism, be it biological or cultural, is to survive.

This integration explains why it is necessary to consider the total culture when analyzing adaptation. Superficially, it might be assumed that attention could be confined to aspects directly related to the environment, such as settlement pattern, material culture, and subsistence. Closer inspection reveals, however, that the amount of time consumed in the acquisition and processing of different foods, their relative importance in the diet, whether the tasks are performed by men or women or both sexes jointly, and other aspects of subsistence behavior have repercussions throughout the culture. Whether analysis begins with religious practices, social organization, or some other sector of a cultural complex, explanation of the content and role of the elements involved will inevitably reveal functional relationships with other categories of behavior that are adaptive. While it remains possible that some cultural traits are truly arbitrary and therefore adaptively neutral, our evidence is still insufficient to confirm this judgment. When tempted to dismiss some feature as of no adaptive value, we would consequently do well to keep in mind a warning by George Gaylord Simpson that "Human judgment is notoriously fallible and perhaps seldom more so than in facile decisions that a character has no adaptive significance because we do not know the use of it" (1953, p. 166).

In the following pages, the cultures of the Camayurá, Jívaro, Kayapó, Sirionó, and Waiwai have been dissected and spread out for inspection. This permits us to compare their structures, to see where they are alike, where they differ, and what is the nature of the differences. Analysis of these similarities and differences in the context of the terra firme environment will provide an indication of some of the complex interrelationships that characterize cultural ecology.

THE CAMAYURA

Location and Environment. At about 12°S. latitude, five major rivers join to form the Xingú. Between this point and their descent from the Brazilian highlands, about 2° farther south, they are slow-moving streams with low banks and broad floodplains. Sandstone uplands with sparse scrub vegetation sharply delimit this part of the Xingú basin on the south and east, providing an ecological barrier to the expansion of occupant tribes. Annual precipitation of 70 inches is concentrated between November and April, when monthly averages reach 8 to 12 inches. In contrast, June, July, and August receive less than one inch of rain monthly. Deer, tapir, capyvara, otter, peccary, monkeys, and smaller mammals are abundant, and the rivers are remarkably rich in fish and turtles. Birds of all kinds are also common.

In recent years, the upper Xingú basin has been occupied by eleven tribes with a total population of only about 750. Six are Carib-speaking, two Arawak-speaking, two Tupi-Guaraní speaking, and one tribe speaks an isolated language (Trumai). The Tupi-Guaraní speaking Camayurá number about 110, equally divided between males and females.

Settlement Pattern. The entire Camayurá tribe is gathered into a single village on a small tributary of the lower Kuluene (Fig. 6). The settlement is on high land about 650 feet from the edge of the floodplain; in the dry season, the river bed is more than half a mile away. Six large houses are arranged in a circle about 330 feet in diameter. The central plaza is clear except for a small rectangular building housing the sacred flutes. Behind the houses are temporary platforms for drying manioc balls. Gardens surround the village, expanding progressively outward as the nearby land becomes unproductive. Some quarter-mile distant is the virgin forest.

The thatched communal houses have parallel sides and rounded ends (Fig. 7). Their size ranges from 55 by 30 feet to 66 by 34 feet, while the ridge rises to about 20 feet. Two tall posts support ends of the ridgepole, while shorter posts set about 26 inches apart frame a vertical wall, about eight feet high. Smaller poles extending from the ridge to a horizontal bar along the upper edge of the wall support the thatch, which is produced by folding long grass over slender poles and fastening these units onto the framework from bottom to top in overlapping rows. An entrance is left at the middle of each side. In cold or stormy weather, or when the sacred flutes are played, it is blocked with a door made of palm fronds. During the dry season, sections of thatch are temporarily removed from the roof to admit light.

Each house is occupied by several related nuclear families, whose hammocks radiate from the roof posts to the walls (Fig. 7). The husband's is slung above that of his wife, who keeps a fire going during the night for warmth. An infant sleeps with its mother; older children hang hammocks adjacent to their parents'. An ordinary house has room for five to eight hammocks at each end, or ten to 16 individuals when one is suspended above another. Personal possessions are kept near the owner's hammock; arrows are stuck into the thatch, while baskets containing ornaments and gourds of piqui oil hang from the rafters. The center is a common area, which in bad weather is used by the women for processing manioc and cooking. Household property and food, including baskets of manioc balls, dried manioc tubers, and dried smoked fish, are stored along the walls.

Dress and Ornament. Both men and women are nude except for a fine fiber cord around the waist. At the center front, the women add a small trianguloid ulurí, made from a specially folded leaf. Both sexes may wear cotton bands below the knee. Typical male adornments include small earplugs, a cotton band on the upper arm, and a string of shell disks around the waist. Women wear necklaces of similar beads. On festive or cere-

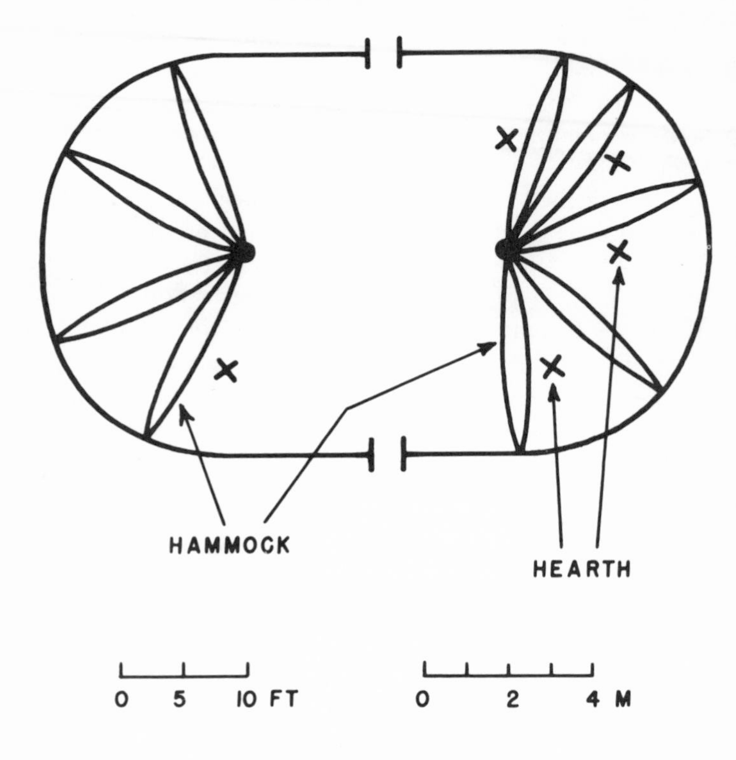

Fig. 7. Ground plan and side view of a Camayurá communal house occupied by several related nuclear families. Hammocks are suspended between roof posts and the wall at both ends, leaving the center free for domestic activities. Fires (X) are kept burning under the hammocks at night for warmth.

monial occasions feather ornaments, palm skirts and capes, and bark ankle wrappings are added. Achiote, a red vegetable paint, is used for body decoration.

SUBSISTENCE. The Camayurá are farmers, and fields are individual property although clearing is done communally by male members of the household. Cutting occurs between June and August so that burning and planting can be scheduled to precede the first rains, which come at the end of September (Fig. 8). Prior to burning, branches are cut up and stacked at the edge of the clearing to provide a reserve of domestic firewood. The earth is hoed into mounds about three feet in diameter and five to 6.5 feet apart for manioc cuttings, nine or ten of which are pushed into one side of each hill. After planting is finished, a bowl of mashed sweet potato is offered to the three manioc guardian spirts. Sweet potatoes are distributed between the manioc hills on the side of the field closest to the village. Maize and peanuts are sown last.

Weeding is done until December, when the plants are well established and the piqui harvest begins. Piqui trees, which produce a large oily fruit, and mangaba trees yielding an apricot-like fruit, are planted around the village. Since they do not bear for ten to 15 years, production usually begins about the time the village is abandoned. Nonfood crops include cotton, tobacco, achiote, gourds, and calabashes. Women plant and pick cotton and help with the harvesting of piqui fruit, sweet potato, and maize; all other agricultural work is done by men.

The staple food, eaten daily the year around, is cassava bread, and processing the bitter manioc is consequently a never-ending task. Since three principal steps are involved, three women usually work together. One peels the tubers with a shell scraper and places them in a pan of water. The second grates the peeled tubers, while the third places about a quart of pulp at a time on a tightly woven mat, which she rolls up and squeezes, allowing the juice and suspended starch to pass through the mesh into a large pottery vessel. When all of the poisonous juice has been expelled, the pulp is pressed into a ball and placed on a basketry tray to dry. After several days of drying, it is ready for use or for storage. The poison can also be eliminated simply by allowing peeled roots to dry before grating, which in this case is deferred until the time of use.

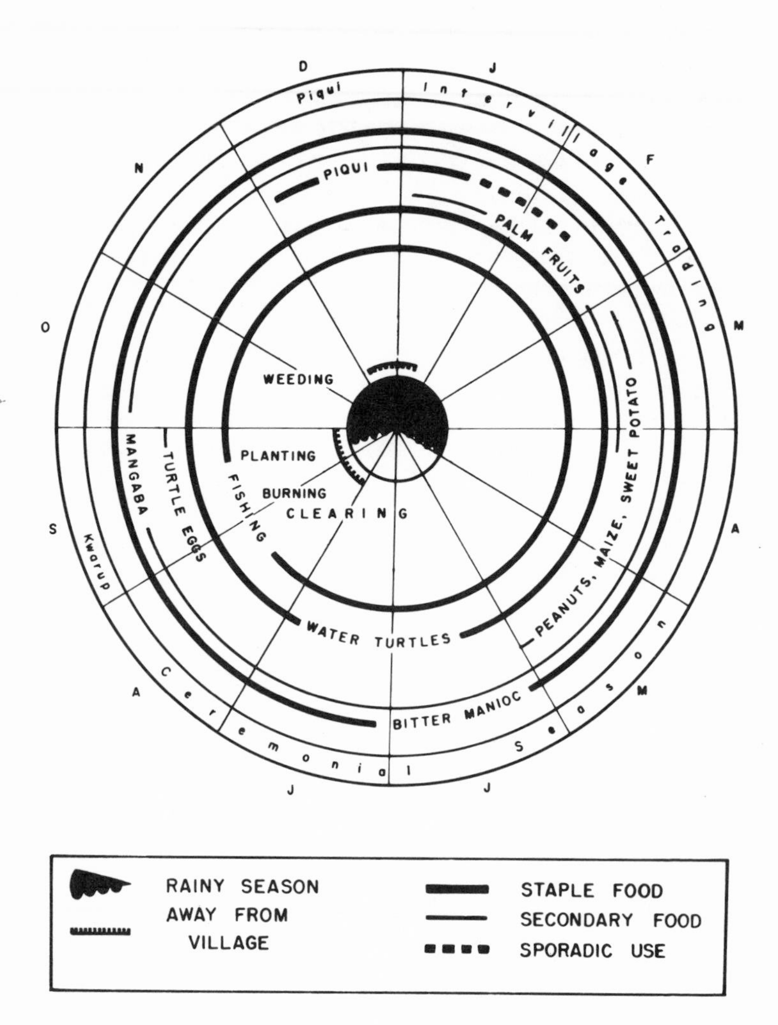

Fig. 8. Annual subsistence round of the Camayurá. Fish, water turtles, and bitter manioc are year-round staples. The village is abandoned twice annually to take advantage of seasonal subsistence resources: turtle eggs at the end of the dry season and piqui fruit during December and January. Other foods are of secondary importance. The marked seasonality of rainfall has led to a concentration of ceremonial activities during the dry months (May through August), and to an emphasis on intervillage trade during the latter part of the rainy season when flooding permits shortcuts by canoe.

All by-products of manioc processing are used. The juice is boiled for hours to eliminate the volatile prussic acid poison, after which it is either drunk or made into soup by adding fish and sweet potatoes. Prior to boiling, the starch may be allowed to settle for extraction. To make cassava bread, the balls of grated pulp are pulverized in a wooden mortar. The resulting flour is sifted, mixed with a little starch, and spread on a shallow circular griddle to bake into a large circular wafer, known as cassava bread, which is often eaten folded around a piece of cooked fish.

During the month of December, piqui fruit assumes a major place in the diet (Fig. 8). The peeled fruit is boiled until the pulp separates from the seeds. Oil rises to the surface and is skimmed off and stored in gourd containers. After cooling, the seeds are removed and dried; the pulp is pressed through a sieve and may be eaten immediately or wrapped in leaves and stored under water, where it keeps for months.

Although game is plentiful, only birds are commonly eaten. Birds are also captured and kept alive for their feathers. Monkey and ocelot are hunted to provide bones needed for the manufacture of arrow points. Wild vegetable foods play relatively little part in the diet, but bocaiuva palm nuts, coco babão, and palmito are eaten in season. Two species of ants are a special delicacy; in one case the head is eaten and in the other the larvae are consumed. Salt is obtained by burning a marsh plant.

Fish are a primary staple. They are so abundant in the lakes that in half a day two men with bows and arrows can fill a canoe to the point of capsizing. One such catch contained 80 fish of 20 different species. Weirs and traps are also employed, and dams are built to create ponds of still water that is poisoned by immersion of crushed vines containing rotenone, which stupifies or kills the fish. Fish are surrounded by numerous taboos; for example, they may not be cooked, eaten, or even touched by a menstruating woman.

From August through September, when the water is low, turtles emerge to lay their eggs on the extensive sand beaches of the upper Xingú. Camayurá families often camp for several weeks on the river bank to subsist on turtles and eggs.

Other Activities. The Camayurá not only have a well defined division of labor along sex lines, but also exhibit incipient occupational specialization. Men make objects used in their subsistence activities, such as bows and arrows, canoes, paddles, fishtraps and other baskets; and carve gourd vessels into spoons, cups, dishes, bailers, and containers. Wooden stools and musical instruments, principally flutes, are also made by men. Women spin and weave cotton, twist twine from burití palm fiber, and weave hammocks. Both sexes manufacture necklaces and combs, in addition to collecting firewood, paddling, and carrying burdens.

The principal weapon is a bow about 6.5 feet long. Four men have become specialists in the manufacture of these bows, and other men secure their bows from these individuals. Arrows are about five feet long and consist of a cane shaft, hardwood foreshaft, and point fashioned from stingray spine, monkey arm bone, or the rib of a tapir or jaguar. A bamboo sliver point is used for tapir and jaguar hunting. These weapons perform with great accuracy up to a distance of about 100 feet.

Other specialized occupations are canoe building, hammock weaving, the manufacture of certain shell ornaments, and shamanism. Except for hammock weaving, all are male activities. A man becomes a shaman during a serious illness, a fate that befalls almost everyone sometime during his life. Initiation takes place during curing. The patient's father or another old man smokes tobacco until he goes into a trance, during which he is able to withdraw the object that has entered the body and caused the sickness. The nature of the object identifies the spirit that will perform services for the new shaman when summoned by tobacco smoke. The initiate is able immediately to perform simple cures with the aid of his supernatural assistant; serious illness requires treatment by several shamans and special rituals, however, since it is believed to result from sorcery. During his life, a shaman may acquire several additional spirit helpers.

Social Organization. Kinship is the basis of Camayurá social relations. Residence is matrilocal until the birth of the first child, after which the couple joins the household of the husband's father. Monogamy predominates but polygyny is permitted, and multiple wives are often sisters. On the death of a husband, his brother assumes responsibility for the widow, permitting her to continue living in the same house. Since kinship is classificatory, a man is free to move from the house of his real brother to another belonging to a classificatory brother if he finds the company more congenial. The man who organized the labor for construction of the communal house assumes the role of house chief; on his death the village council choses his successor from his sons or younger brothers. The household constitutes a cooperative work group for gardening, fishing, manioc processing, and other activities, and members also go visiting together.

The village is governed by a council composed of the tribal chief, the house chiefs, and other mature men, which meets nightly around a fire at the center of the plaza. The tribal chief is selected by council members from the male relatives of the previous chief on the basis of knowledge of custom and ritual, ability as an arbitrator in disputes, and skill as a shaman. In addition to implementing decisions reached in the council, the chief's responsibilities include organization of community activities, such as garden clearing, fish poisoning, trading, moving to a new village site, and receiving visitors from other tribes. He does not serve as a leader in warfare. His position entitles him to respect, but neither frees him from normal male duties nor brings him any material gain in the form of tribute. Only at death is his unusual status recognized in special burial practices.

LIFE CYCLE. During a wife's pregnancy, both prospective parents observe dietary restrictions to promote the welfare of the unborn child. Although children are greatly desired, abortion may be attempted if the previous child is still nursing. Birth takes place in public in the house, and for a month thereafter the father must remain indoors. While he is immobilized, relatives supply the family with food. If the infant is deformed, or if twins are born, or if the previous child is still nursing, infanticide is practiced. The offspring of an unmarried woman is also destroyed.

A child is rarely separated from its mother until weaning, which occurs between the ages of three and four. Up to the age of six, it is allowed freedom to sleep, eat, and play when and where it chooses, but is continuously watched by older siblings or adults. Boys from six onward begin to practice marksmanship with miniature bows and arrows, and may accompany older male relatives on fishing trips. Girls begin a little later to help with household tasks, such as carrying water, processing manioc, and taking care of younger children. They also begin to assist in gardening and to learn the secular dances. Until puberty, both sexes remain under the control of their mother, who disciplines them for disobedience or lack of respect to their elders.

At puberty, when a child has become adept at adult activities, he or she enters a period of strict discipline and formal training that helps to emphasize the transition between carefree childhood and the serious responsibilities of adult life. At first menstruation, a girl is secluded for three or four months behind a screen in the house, where she passes the time spinning, and where older women visit her to teach her the medicines used for abortion, what to do during pregnancy, and how to treat her future husband. Boys are similarly secluded about the age of 14, and kept busy making bows and arrows and other objects while they are instructed by their fathers or other old men in the tribal history, aspects of ritual and belief, how to play the flutes, how to behave in a raid, the importance of continence, and other essentials of the adult male role. Both sexes are subjected to scarification to increase their tolerance for pain. Seclusion ends in September, at the time of the kwarúp ceremony, when marriages generally take place.

Marriages are arranged by the parents while the children are still young. Actual or classificatory cross-cousins are preferred spouses. Intertribal marriages occur, however, either with captives or because of the absence of a suitable mate within the tribe. Although the foreigner is given kinship status, the children are considered only half Camayurá. Nine of the 40 adult women and six of the 41 adult men comprising the Camayurá population in 1949 were from other tribes, five of the women and two of the men having been taken as captives. Several Waurá women had been sought because of their knowledge of pottery-making. The marriage ceremony consists in the couple cutting one another's hair, which has been allowed to grow during puberty seclusion. Marriage does not free a man from his father's domination; on the contrary, sons stand in fear of their father's anger throughout his life and obey his wishes. Adultery is not considered a serious crime, but custom dictates that the offending partner should be beaten. Both men and women advance in prestige by fulfilling their obligations cheerfully and competently and performing their allotted tasks with skill. Men derive their greatest renown, however, not from skill in fishing or gardening, but from becoming a champion wrestler or spear thrower in competitive games.

The Camayurá recognize that death may come as a result of accident or sorcery, but the fundamental cause is believed to be desertion by the spirits. The deceased is wrapped in a hammock and buried in the village plaza. The position of the body is governed by the individual's age and status. For a chief, a deep shaft is dug with an alcove at the base, in which the hammock is suspended between two posts. The immediate relatives cut their hair, scarify their arms, and wail loudly. After the burial, they remain secluded behind a screen in the house until the next kwarúp ceremony.

CEREMONIES. Camayurá ceremonial occasions are fixed by the calendar cycle; and since the major ceremonies involve the participation of members of other tribes, they afford opportunities for trading, sport competitions, visiting, and feasting as well as ritual expression. The two great festivals are the kwarúp dance cycle at the beginning of the rainy season in September, and the yawarí ceremonial at the beginning of the dry season in late April. The emphasis on these two occasions is somewhat different.

The kwarúp ceremonial, which lasts several days, begins with a symbolic reenactment of the tribal origin myth. This annual performance is believed to be essential to the perpetuation of the tribe. As a concrete expression of its significance, the beginning of the kwarúp occasions the release of boys and girls from puberty seclusion and the solemnization of marriages, which have long since been arranged. Special dances in this cycle promote the growth of plants and the inception of the annual rains; others honor specific tree spirits.

The yawarí is a secular festival and is always attended by one of the neighboring tribes. The main event is a contest between the men of the two tribes, in which members of one team attempt to strike their opponents in the legs with blunt-ended spears thrown with a spearthrower. Wrestling matches are another form of competition. Young men train for these events and observe dietary and sexual restrictions prior to each contest; rules of the game are strictly observed, and great prestige accrues to the individual who becomes tribal champion. Large quantities of food are prepared for the accompanying feast and for presentation to the visitors on their departure. The more abundant the food, the higher the reputation of the host village.

Two ceremonies of shorter duration promote the abundance of two staple foods. Men, women, and children participate in a dance in honor of the piqui spirit, which is held in December while the tribe is camped at the former village where the harvest takes place. At the beginning of the dry season in April, a series of masked dances is performed by the men to assure an abundance of fish during the coming months. Women and children are excluded and must remain behind closed doors in the house.

TRADE. The existence of specialists in the manufacture of certain articles within the village has led to the emergence of a kind of market, which takes place whenever a house group accumulates a surplus of articles suitable for exchange. An announcement is made in the evening in the village plaza, and the following morning the objects are placed where they can be inspected by potential buyers. When a person sees something he wants, he goes to his house to select an object or objects for exchange, which he places in front of the item of his choice. If the seller agrees, the deal is closed. There is no discussion or haggling. If a member of another household has something to "put on the market," he is free to place it in the seller's line.

Intervillage trading operates in a similar fashion, usually taking place during the rainy season when high water permits more direct routes to be followed between villages. Most of the upper Xingú tribes have become specialists in the manufacture of certain types of articles. All pottery used by the Camayurá, for example, is manufactured by Waurá women; and since pottery vessels are essential for processing manioc and piqui as well as for other types of cooking, and are susceptible to breakage, a relatively steady supply is required. Flutes, utilized in many of the ceremonies, are obtained from the Mehinácu, while other tribes specialize in making shell necklaces or canoes. The Camayurá are the best bow makers. Intertribal exchange is based on a recognized scale of values in which, for example, a snail shell necklace is worth two bows. Food also is traded, but is lowest on the relative scale of values. One tribe, the Avetí, has specialized in trading; members travel from village to village exchanging produce and also obtain articles of European origin for trade to other tribes.

WARFARE. The Camayurá are on peaceful terms with other tribes of the upper Xingú basin, although the relationships are not relaxed. In spite of continual intertribal trade, sport competitions, and participation in ceremonies, emotions are tense and strict etiquette is observed during visits. Visitors never sleep in the host village, but camp nearby and leave as soon as possible. The kinship bonds resulting from intertribal marriage and the release provided by vigorous athletic competitions help to dissipate tensions that would otherwise find expression in outbreaks of violence.

A state of permanent hostility exists between the Camayurá and tribes outside the Xingú basin. Raids occur sporadically to capture women or to avenge deaths inflicted during a previous raid. Captives are adopted. Although it is difficult to judge the extent of warfare prior to European contact, it appears to have been less intense than among the Jívaro and other tribes of the western Amazon basin.

RELIGION AND MAGIC. Along with the sun, the moon, and many elements of their culture, the creator gave spirits to the Camayurá. The spirits are audible and visible only to shamans, who describe them as dwarfs, birds, insects, or animals. They live in the forest and the air, and are dedicated to preserving the health of the Camayurá tribe and promoting the growth of their subsistence animals and plants. Powerful spirits, who can be summoned for help when needed, also inhabit five kinds of sacred objects: two kinds of sacred flutes, a special gourd rattle, a bullroarer, and a wooden mask. All are kept in the flute house when not in use and must never be seen by women. The impersonation of spirits and the ceremonial reenactment of tribal myths are believed necessary to assure the perpetuation of the Camayurá economic and social order.

Sorcery is not caused by spirits, who are never pictured as evil. It is accomplished instead by the injection of a foreign object into the body, and is combated by removing this object by sucking and by smoking tobacco. Tobacco has a powerful spirit, able to summon other spirits to the aid of the shaman. Although sorcery is not practiced by Camayurá, it is greatly feared. This latent danger produces an undercurrent of tension in their relations with other tribes.

THE JIVARO

LOCATION AND ENVIRONMENT. The tropical rain forest reaches its western limit on the lower slopes of the Andes, at an elevation of about 5000 feet. Rainfall here averages 80 to 100 inches annually, and although monthly totals are variable it is a rare month that receives less than 6 inches. Innumerable streams cascade down the mountain slopes and join to form increasingly larger rivers as they enter the lowlands. The combination of continual moisture and warmth produces a lush vegetation, which in turn supports a rich terrestrial fauna. Fish are few in the swift and rocky streams on the west of the area, but become more numerous as the current grows more placid.

The Jívaro occupy an area of some 25,000 square miles, bounded by the mountain slope on the west, the Rio Pastaza on the east, and the Rio Marañon on the south (Fig. 6). They share a common culture and speak the same language, but have no social or political unity. On the contrary, intense feuding and warfare between local groups continues in spite of recent efforts at pacification and missionization. The present population has been estimated at about 20,000, with females outnumbering males by more than two to one.

SETTLEMENT PATTERN. A Jívaro village consists of a single communal house occupied by a patrilineal extended family. It is located in a defensible place, such as on a hilltop, at the bend of a stream, or (if the terrain is more rugged) on a rocky ledge. Although accessible to a small stream, the house is erected some distance from the bank to minimize the likelihood that passers-by will detect it. Access paths are as poorly defined as possible and provided with spring traps and deadfalls to discourage unwelcome visitors.

The house is 50 to 100 feet long, with straight sides and rounded ends (Fig. 9). Walls are solidly constructed of closely spaced vertical poles, and the roof is thatched. A tall narrow door at each end is securely closed at night with planks. The interior is divided into male and female halves, but with no formal partition between them. Platforms about four feet from head to foot, five feet across, and raised 12 to 16 inches above the floor are built along the wall of the men's section to serve as beds. The uprights at the corners continue to a height of about 6.5 feet, where another platform is constructed for storage. A horizontal bar 8 to 12 inches from the end serves as a foot rest, and a fire is kept burning beneath it on chilly nights. Lances, blowguns, looms, and other kinds of male property are arranged along the walls, while smaller objects are stored in baskets or nets that hang from hooks

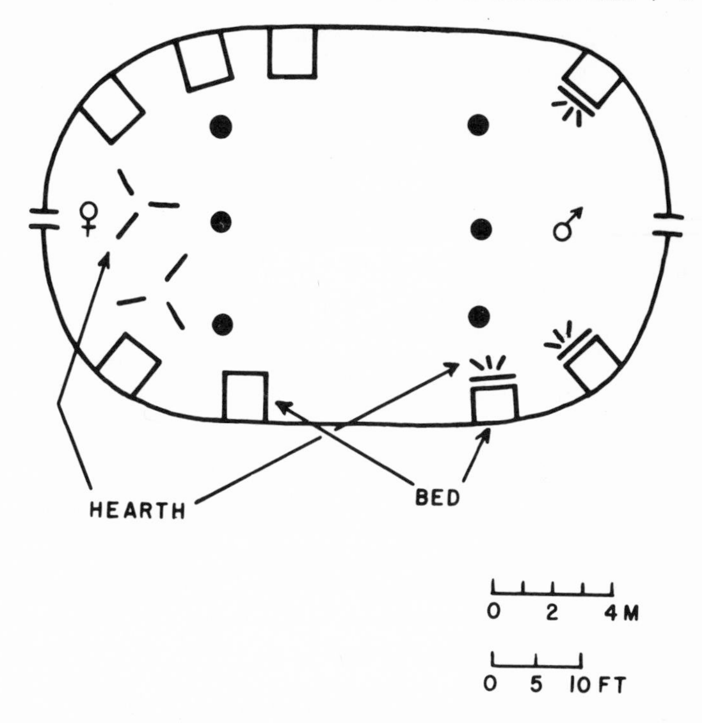

Fig. 9. Ground plan and side view of a Jívaro communal house inhabited by a patrilocal extended family. The men occupy the right half and the women and small children the left half, each accessible by a separate entrance. Men's platform beds have a foot rest, beneath which a small fire burns at night for warmth. Women's beds are longer and walled for privacy. Cooking fires fed by three large logs occupy the center of the women's quarters.

suspended from the rafters or are laid on the shelf over the bed. A large signal drum is hung horizontally near the door.

The women's portion of the house is similarly arranged. The platform beds that line the walls are longer than the men's, however, eliminating the need for a footrest, and have side walls and a curtained end for privacy. Cooking fires are located near the entrance. Like all Jívaro fires, they are fed by three logs that are moved progressively inward as the ends are consumed. Pottery vessels of various sizes, gourd water bottles and bowls, and other domestic utensils stand on the floor or are stored on hanging racks. Tied to the foot of each bed are two or more dogs.

A recent census of 17 Jívaro households showed them to consist of three to five married men, four to 18 married women, and six to 32 preadolescent children. The population per house varied between 15 and 46.

Dress and Ornament. The principal article of dress is a rectangular piece of cotton cloth, which men wear as a wrap-around skirt extending from the waist to just below the knee. Women fashion a dress from a larger rectangle by passing the upper edge below the left arm and pinning it over the right shoulder. When bloused out above a cord belt, the skirt extends to the knee. Typical ornaments include a headband of fur, feathers, or cotton cords; necklaces, collars, and other ornaments of seeds, shells, teeth, bird bones, etc.; and bamboo tubes decorated with white down and toucan tufts, which are inserted through perforations in the ear lobes. Females wear a small cane plug in their lower lip. Red and black paint and facial tattooing are other methods of adornment.

Subsistence. The Jívaro staple is sweet manioc, and a large garden is cleared at the time a new house is constructed. One or more additional fields are made in the nearby forest when production of the first field declines or when additional food is required for a special feast. The men cut the vegetation and burn it, while planting, weeding, harvesting, and garden magic are primarily tasks of the women. Men plant maize, cotton, and barbasco (fish poison), however, and observe dietary restrictions for the successful growth and maturation of these crops. Subsistence staples in addition to sweet manioc include chonta or peach palm fruit, sweet potatoes, squash, papaya, and peanuts (Fig. 10).

Bananas, plantains, and sugar cane have been added recently. Among other important nonfood cultigens are tobacco, achiote (dye), and several narcotics.

At the conclusion of the manioc planting, a festival is held on

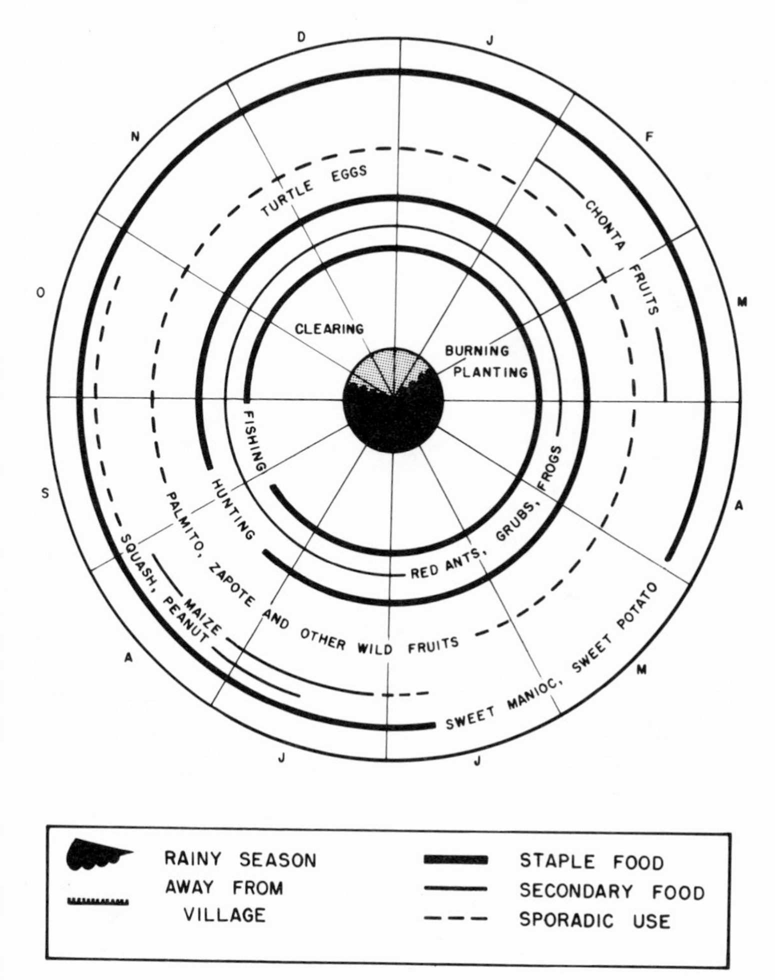

Fig. 10. Annual subsistence round of the Jívaro. Fish, game, sweet manioc, and sweet potatoes are year-round staples. Chonta fruits and maize are important during a few months. Only sporadic use is made of wild plant foods.

five successive nights, during which the women of the household dance and chant to promote the growth of the plants and to protect them from damage. Incantations, magic, and dietary restrictions are employed on various occasions for the same purposes, particularly by the women, who are believed to have a special relationship with the plants.

Most of the women go daily to the garden to tend the plants and to harvest food. Another constant and time-consuming female activity is the processing of sweet manioc into a fermented drink, several quarts of which are consumed daily by each adult. The tuber must be boiled until it is soft. Then half the quantity is mashed and the remainder is chewed and spit into a bowl. The mixture of these two products, which resembles mashed potatoes in consistency, is allowed to stand for a day to ferment. Prior to consumption, a couple of handfuls are placed in a calabash with water, which permits the coarser fiber to be removed when it floats to the surface. A similar fermented drink is prepared from chonta fruit. Manioc leaves are boiled and eaten as a green vegetable.

A combination of three factors—a relative abundance of game, omnivorous taste, and great skill on the part of the hunter—assures the Jívaro household of a steady supply of meat. One or more of the men hunt daily, and their intimate knowledge of the environment and habits of the fauna, combined with a varied repertoire of methods (including imitation of calls, construction of lures and traps, and skill with the blowgun and lance) usually produces results within a couple of hours. When the trail of a band of peccaries is discovered, a communal hunt is organized and the meat divided among the group. Dogs are specially bred and trained for hunting, and one of the major Jívaro feasts is that given when a dog is full-grown. During the three-day event, magic is practiced to give the animal strength, endurance, and a keen sense of smell. Dogs are required to observe certain food taboos to enhance their efficacy as hunters and to preserve their health and wellbeing.

Deer and tapir are not eaten because of their supernatural qualities, but all other kinds of game animals and birds are consumed, including monkeys, sloths, capyvaras, agoutis, anteaters, caymans, and birds—even very small ones. Snakes, beasts and birds of prey, vultures, and most nocturnal birds are not eaten.

Women never hunt, but may accompany their husband to carry home the game.

Several methods of fishing are employed, depending on circumstances. They include diving and catching with bare hands, harpooning, hook and line, trapping, netting, and poisoning. The latter is a community activity; otherwise fishing is done by men and boys.

Food gathering provides a relatively unimportant part of the Jívaro diet. Palmito is the principal wild plant consumed, and it is utilized mostly during traveling or during periods when taboos on other foods are being observed. Like other northwest Amazonian tribes, the Jívaro eat a number of insects. Especially favored are a red ant and the large white grub of the chonta beetle, whose flavor has been compared to pork sausage spiced with nutmeg. Turtle eggs are another delicacy, exploited seasonally. Numerous forest plants supply the fibers, woods, resins, and other raw materials for the manufacture of most articles of material culture, in addition to medicines and magical substances.

OTHER ACTIVITIES. A sexual division of labor prevails in manufacturing activities. Men perform all tasks involving wood, such as house, bed, and loom construction; canoe making; and carving paddles, drums, lances, blowguns, darts, digging sticks, and shields. They also weave baskets, spin and weave cotton, and make many of the ornaments they wear. Women have special affinities with the earth and are consequently assigned earth-related activities, including agriculture, pottery-making, and cotton-dyeing. A large proportion of the daily household work also falls to them, such as cooking, preparing drinks, caring for children and dogs, and carrying burdens and water. Men make fire and bring the logs used as fuel, since both tasks involve wood.

The only specialist is the shaman. His principal function is curing, but through his control over the spirit world he can also influence weather, cause death or illness, and identify the culprit in case of death. His knowledge is acquired during training under an experienced practitioner. While he is an apprentice, he must observe strict dietary and sexual taboos or suffer the punishment of death. A shaman's status is somewhat diluted by the fact that all old men and women, whether practitioners or not, have special powers over the spirit world, and he receives little material advantage from his role. On the contrary, since the Jívaro look upon the death of any adult as an unnatural event for which sorcery is responsible, and since all shamans are sorcerers, he becomes a primary target for blood revenge.

Social Organization. Although each Jívaro household is an economically and politically independent village, it is part of a larger social group composed of about half a dozen houses occupied by related families scattered over a distance of some nine miles along both sides of a small stream (Fig. 11). This local community has no chief except in time of war, when one is elected from among the participants. Members of the different households visit back and forth almost daily and join in feasts on ceremonial occasions. Individual households are exogamous and the local group is generally endogamous. Kinship ties do not eliminate the danger of sorcery, however, so that a chance meeting in the forest or even a social call involves a potential threat. Visitors consequently announce their arrival from a distance with a warning whistle, and the visit begins with a stereotyped ex-

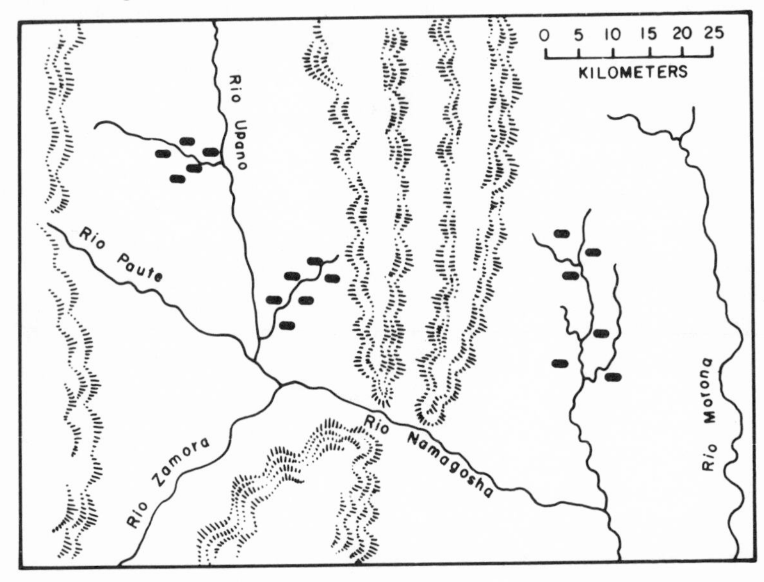

Fig. 11. Geographical distribution of three Jívaro communities showing the tendency for five or six households to cluster along a stream. These groups of houses shelter related families and constitute the largest stable social grouping in Jívaro society. Relations between members of different clusters typically are hostile (after Danielssen, 1949, Map II).

change between visitor and host that lasts nearly half an hour, during which feelings of tension dissipate so that normal social interaction can begin. Relations of latent or active hostility prevail with other local communities, the nearest of which may be less than 20 miles away.

The highest permanent chief in Jívaro society is the male head of a household. Although his power is theoretically absolute, an atmosphere of cooperation and mutual respect prevails between him and his wives, who are essential to whatever status he may hope to achieve and crucial to the wellbeing of the family. Women are important not only because of their tangible contributions but because of their supernatural influence:

> Only a woman can act upon the female spirit of the manioc plant so as to make it produce an abundant crop; cultivated by a man it would yield but a meagre harvest. Only under a woman's care will the swine and the fowls increase and thrive, and the hunting-dogs become able to trace the game. With no education by a prudent and skillful mother, the sons and daughters would hardly grow up brave men and intelligent family-mothers. The man who does not own a good wife, properly initiated during the "feast of women," will soon find himself deprived of the necessities of life, nay, will see his whole household ruined, however well he may have been provided in the beginning with fields and domestic animals (Karsten, 1935, p. 256).

LIFE CYCLE. Although contraceptives and abortives are known, they are seldom employed because children of either sex are desired. From the time that pregnancy is advanced until the child ceases suckling, both parents observe a variety of dietary taboos and other restrictions, including sexual continence. Infanticide is practiced if the infant is deformed or if the father is from another tribe, but twins are not killed. In order to promote the spiritual strength of the child, the mother rests in the house for three days after the birth. The father remains inactive for eight days, being especially careful not to hunt or to work with a machete during this period. Young children enjoy great freedom, but about the age of seven begin to learn adult tasks. Boys accompany their fathers in hunting and warfare, and are harangued frequently to arouse hatred for the enemy and a thirst for vengeance. Although they do not actively engage in fighting before maturity, they become familiar with and inured to bloodshed, pain, and violent death.

At puberty both boys and girls are subjected to special restrictions in diet and activities. The boys' initiation feast, held about the age of 15, is a five-day event attended by members of other villages. Henceforth, the boy resides in the men's side of the house, and marriage follows shortly after. Girls often marry much younger, but sexual relations are delayed until after puberty. Village exogamy is practiced, and marriages with cross-cousins are preferred. Residence is patrilocal, but when a man has accumulated several wives and a number of children he may found an independent household. Co-wives are often sisters, but need not be. Although the age difference between first and last wives may be considerable, co-wives get along well together, dividing the work and displaying no rivalry or jealousy.

Old people enjoy high status whether or not they remain physically able to work. All the great feasts must be led by an old man or woman if their ritual aspects are to be successful. Death is a sad event, in which grief for the deceased is combined with hostility toward the person responsible for the death. Wives cut their hair and refrain from painting their bodies or putting on their usual ornaments. If the deceased was head of the household, his body is placed in a hollowed log, along with certain possessions, and is suspended from the house rafters, after which the house is abandoned. Other adults are buried beneath the house floor or in the vicinity. Property not intimately associated with the deceased, such as his best loincloth, blowgun, and extra lances, are inherited by his sons (or in the case of a woman, by her daughters). Wives of the deceased become the responsibility of his brother.

Ceremonies. The Jívaro hold great feasts on five occasions: for boys at puberty, for girls at the time of marriage, for dogs when their training is completed, for warriors who have taken a head, and for young children. Members of other nearby villages are invited to participate in the ceremonies, feasting, and dancing, which last five or six days. These rituals not only impart strength and long life to the honored individuals, but insure abundant harvest, plentiful game, and success in all aspects of life from subsistence activities to domestic relations. The feast terminating the period of ceremonies and restrictions that follows the taking of a head requires lengthy preparation, including planting extra fields and acquiring large amounts of meat and fish to feed the participants, as well as the manufacture of pottery vessels, stools, and other objects to be utilized by the guests. A man who does not fulfill his obligations to hold one of these ceremonies when the appropriate occasion arises will not only suffer loss of prestige but will seriously endanger the wellbeing of his family by incurring the displeasure of the spirits.

TRADE. Their constant state of mutual enmity does not prevent trading between Jívaro subtribes. Certain kinds of objects, such as salt and the palm used for arrow shafts, are not available everywhere in the region. Other traded items are blowguns, arrow poison, small drums, and hunting dogs. In recent times, the Jívaro have obtained pigs, chickens, axes, knives, guns, ammunition, and cloth from their civilized neighbors in return for salt, deer meat, and blowguns. Trading is done on an individual basis; and a person traveling through enemy territory on a trading mission runs the risk of becoming a target for blood revenge.

WARFARE. Active hostility may be provoked by a variety of incidents, such as a series of deaths that indicate sorcery, the abduction of a woman, or the murder of a member of the group while traveling through foreign territory. Messages are sent to the men in neighboring households to solicit their participation. The group elects a leader, who exercises considerable authority over the organization and strategy of the expedition and is obeyed by the participants. Preparations include divination to learn the most propitious time for attack and the probable outcome, sending spies to learn the lay of the land and the defensive measures that may have been taken by the enemy, and, on the night before departure, a war dance.

The aim of warfare is annihilation of the entire enemy household. Men, children, and old women are killed, and young women are spared only if they attempt no defense. The house and contents are burned, the dogs killed, the gardens uprooted. Heads of adults who are not related to a member of the war party are taken as trophies, after which the victors retreat as quickly as possible. When they feel safe from pursuit, they erect a temporary shelter and undertake the shrinking of the heads, a process that requires about 20 hours. The party then continues home, where the victory celebration begins.

RELIGION AND MAGIC. The Jívaro envision the world as containing spirits that inhabit people, animals, plants, striking geographical features (such as hills, rapids, and volcanoes), and even objects of human manufacture. These spirits may move from one "home" to another, and the spirits of human beings often enter certain animals, especially the deer and the tapir. Spirits may also exist in disembodied form. The vast majority are evil or at least potentially dangerous, and care is taken not to arouse their anger. The souls of shamans, sorcerers, and enemies killed in warfare or revenge are particularly powerful and malignant. Fortunately, they fear loud noises, so that shouting, rattling shields, and other kinds of racket can be employed to drive them off. Chonta wood is also a powerful protector.

In addition to the spirit world, there are two important deities: the Earth-mother and her husband. One or the other is invoked at various times during the growing season by incantations soliciting their good influence on the maturing plants. The Earth-mother is particularly important because she appears to women who are under the influence of a narcotic and instructs them in the care of plants and animals. It is she who originally taught agriculture to the women.

The Jívaro attribute the death of an adult to sorcery, for which someone must be held responsible. The shaman calls upon his supernatural contacts to identify the culprit, who frequently turns out to be another shaman. The guilty party may even have been a friend or relative of the deceased. Regardless of his identity, the sons and brothers of the victim are obligated to put the culprit to death. If this is impractical, however, a close male or female relative can be substituted. Such blood revenge is a personal matter and does not involve either the taking of a head or the celebration of a victory feast.

THE KAYAPO

Location and Environment. The region between the Rio Araguaya and the middle Rio Xingú consists of forest-bordered lakes and streams interspersed by tongues of savanna, which represent penetrations of the arid uplands that delimit the southeastern part of the Amazon basin. The average annual rainfall of 70 inches has a marked seasonal distribution; during the months of June, July, and August, there is little precipitation. Hunting and gathering resources are abundant, with the Brazil nut one of the principal forest products.

The northern Kayapó, who inhabit this area (Fig. 6), are a group of subtribes that speak a Ge language and share a similar cultural pattern, but exist in a state of mutual hostility. In 1897 they were distributed over an area of some 40,000 square miles and had a population estimated at about 5,000. Since the villages tend to be located away from major rivers, where they would be more readily encountered, it is possible that the actual total was significantly higher. As recently as 1954, a new village with more than 500 inhabitants was discovered in the forest 25 miles from the left bank of the Rio Xingú. Beginning with European contact and accelerating during this century, the Kayapó have suffered severe decimation both from disease and as a result of hostile encounters with Brazil nut collectors and rubber gatherers. Entire villages have become extinct and others have been so reduced in size that many important social institutions can no longer be maintained. Census figures on two groups indicate a slight predominance of females.

Settlement Pattern. The Kayapó alternate between settled village life during the rainy season and a wandering existence during the dry months of the year. Community size is variable and aboriginal figures are difficult to obtain, but the normal range appears to have been between about 200 and 700 people, with a few reports of villages exceeding 1,000 in population. During the dry season, the entire band may move together or it may split into smaller groups, leaving the ill and infirm behind in the village in the care of relatives. There seems to be no well-defined territory, nor is there any specific information on the distance separating villages. Since wandering takes place at a distance of three to five days travel from the village and accidental encounters with other groups are rare, their spacing must be considerable.

During its nomadic phase, the group follows a well-defined routine. Camp is broken at dawn and the march to a new site two or three hours distant is led by the warriors. A new camp is set up in a very short time, with each person doing a specified job according to sex and age. Huts are erected by the women and arranged to duplicate the village plan. The daily round of activities otherwise differs little from that in the village itself.

The village, which may serve as a base of operation for an indefinite period, is located on a well-drained spot near a small stream, which serves as a source of water and a place for bathing. A number of communal houses are arranged in a ring around a plaza, from which paths lead outward to the streambank, the gardens, and the forest (Fig. 12). If the population is large, there may be two concentric rings. At the center of the plaza or at one side is a men's house. This is the only public structure other than a community oven, which serves as a social center for the women.

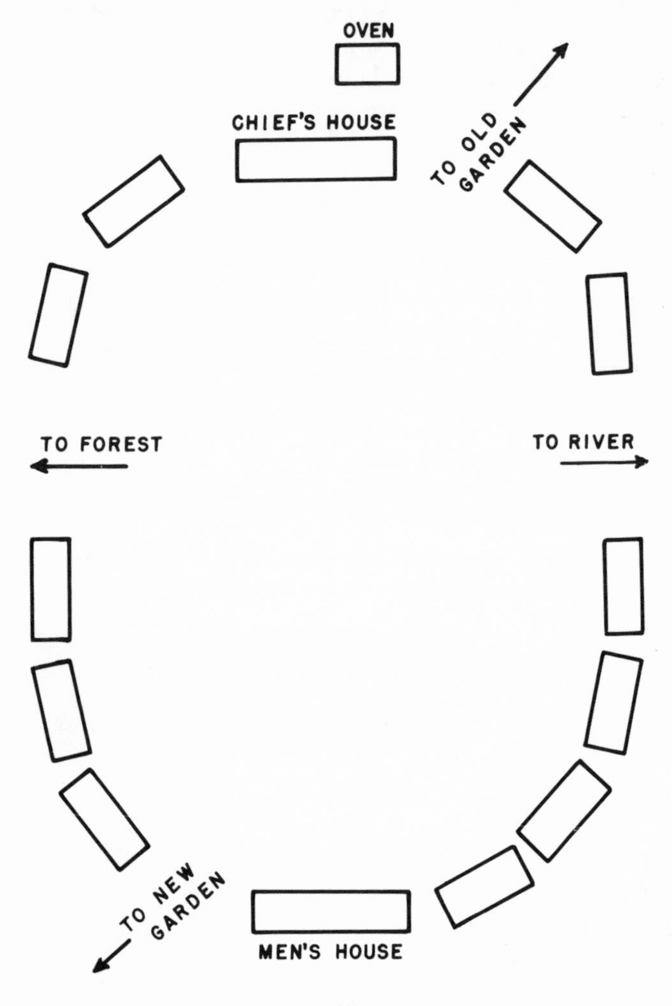

Fig. 12. Plan of a Kayapó village. Communal houses, each occupied by a matrilocal extended family, are arranged around a plaza. Paths lead in four directions: to the river, the gardens, and the forest (after Frikel, 1968, Fig. 2).

The houses are rectangular, averaging about 26 feet long by 13 feet wide. The vertical wall is about 6.5 feet high, and the ridge rises another 6 or 7 feet. The framework of posts and poles is covered with palm leaf thatch, except for the wall toward the plaza, which is only partly screened by the overhang of the roof thatch. A door is left in one or two of the other walls to facilitate access to the interior. The men's house and the chief's house are similar but larger structures.

The dwellings have no partitions and no permanent features except cooking fires, one of which exists in each nuclear family area. The Kayapó do not use hammocks; they sleep either on freshly cut palm leaves, mats, or in some cases, on low platform beds. Household utensils and women's possessions and ornaments are hung from the walls or rafters.

Dress and Ornament. No clothing is worn by either sex, and few ornaments are used except on festive occasions. Black painted designs are applied all over the body of children, who also wear ear ornaments. During dances and ceremonies, special costumes, feather and bead headdresses, necklaces, and other decorations are worn by both sexes. Men also wear lip ornaments.

Subsistence. Until recently, the Kayapó depended more upon wild foods than upon cultivated plants (Fig. 13). Aboriginally, the staple crops were maize and sweet potatoes, followed by sweet manioc and yams. Cotton, tobacco, and achiote are also planted. Under the restrictive pressures of European encroachment, fields have been enlarged and new crops, such as rice, sugar cane, and bitter manioc, have been adopted.

The location of a new field is chosen by the chief after listening to suggestions made in the nightly council. Clearing is done by the men and boys before the dry season is advanced, to allow time for drying before burning. The plot is divided into family areas, and planting and weeding are done by nuclear family groups. The staple crops are harvested by women. During the first and second years, the field is planted in maize and sweet manioc. In the third year, sweet potatoes are substituted, and since the plants continue to bear for several years, the area in sweet potato production is always relatively large.

The traditional dish of the Kayapó is a kind of bread made of maize, sweet manioc, or sweet potato dough, sometimes with bits of Brazil nut, meat, or fish included. This is wrapped in banana

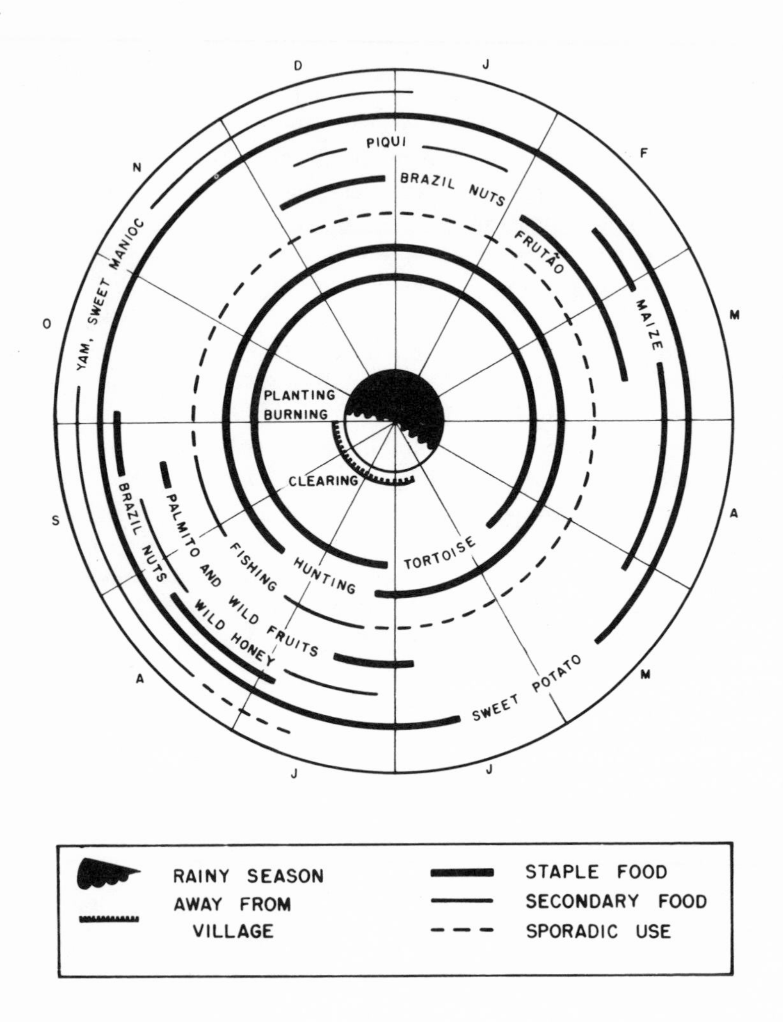

Fig. 13. Annual subsistence round of the Kayapó. Game, land tortoises, and sweet potatoes are year-round staples, while Brazil nuts and maize are primary foods during certain months of the year. During the dry season (July to September), the village is abandoned except by the ill and infirm, and the population divides into extended family bands to wander through the forest subsisting on fish, palmito, Brazil nuts, and wild fruits.

leaves and baked in an earth oven covered with hot stones and dirt. Chunks of meat and palmito are also baked in this fashion. Fish, maize ears, and sweet potatoes are roasted over an open fire. Certain kinds of fruits are pulverized and mixed with water. Brazil nuts are either eaten raw or grated to produce a milky beverage. No fermented foods or drinks are prepared.

Hunting is a sport as well as a subsistence activity, and men are expert at tracking and at imitating animal and bird cries. The head of a family may go off hunting for several days alone, but more commonly several related men and boys join together. Communal hunts are organized for peccary and tapir, and also when a large quantity of meat is required for a festival. Dogs are scarified on the head and back with jaguar claws and fed meat before the hunt to make them more aggressive. When in the village, however, dogs must scavenge their own food. Although the bow and arrow is used for hunting birds and to wound and thus reduce the speed of other game, the traditional weapon is a wooden club. Small animals are divided among relatives, with shares being sent to the chief and to the mother or sister of the hunter. Large animals, such as deer, tapir, and peccary, are delivered to the chief for distribution.

Fishing is more important during the dry season, when the lowered water makes it more productive. The time and place of a fish poisoning venture is decided in the men's council. A division of labor is assigned, by which one group of men cuts the poisonous vines while another builds a stone barricade across the stream to reduce the speed of the water. The catch is seldom abundant, and many streams can be poisoned profitably only once a year. The bow and arrow is used for individual fishing, and the hook and line has been recently adopted. Children are sometimes sent to catch fish to feed sick relatives.

Collecting is a primary source of food, particularly during the dry season. Innumerable plants are exploited, the majority of them palms. Palmito or palm cabbage is eaten either raw or baked. Another mainstay is the Brazil nut. Although nuts ripen late in the dry season, they do not fall to the ground until the stems have been weakened by heavy rains. Dry season exploitation thus depends on climbing the trees, a dangerous activity allocated to boys. Fallen nuts are collected and carried to the village, where they are stored for subsequent use. Tortoises are also gathered and are kept alive until needed. They are very abundant and some 2,000 per year can be obtained within a few hours' radius of the village.

OTHER ACTIVITIES. The division of labor is more pronounced in arts and crafts than in subsistence activities. Most of the essential tools and utensils are made by men, but not all the items utilized by other Amazonian tribes are produced. There are no hammocks, cotton cloth, canoes, or paddles, and only a few musical instruments. Baskets, twine, bows and arrows, clubs, and ornaments are the principal manufactures. No pottery is made, and women have few craft specialties aside from spinning cotton and making certain kinds of seed and cotton ornaments. Most of their time is spent in child care, food preparation, gardening, food gathering, fetching water, and collecting firewood. Although men construct the houses in the village, women build the shelter for the community oven and usually erect the huts in the dry season camps.

The traditional weapon among the Kayapó is a wooden club, which is used in hunting, warfare, and ceremonies. It is circular in cross-section, cylindrical or slightly expanded toward one end, and ranges from two or five feet in length. The shorter clubs are used in hunting, the longer ones for other activities. They are wielded with a rapid movement and sufficient force to produce a whistling sound. In warfare, the head of the victim is the target, but during sporting contests combatants aim at the shoulders and upper arms, desisting only when too exhausted to continue.

According to oral traditions the bow and arrow is a relatively recent introduction. Bows are about 6.5 feet long; arrow length is four to 5.5 feet. A wood, cane, bone, or (in recent times) a metal point is inserted in a cane shaft, sometimes with a wooden foreshaft. Poison is not employed.

The only specialized occupation is shamanism, and there may be only two or three shamans in a population of 200. The shaman is an old man or woman who has established communion with the soul of a deceased person, an animal spirit, or a supernatural monster. Ordinary treatment of illness, which involves administration of plant remedies, painting with achiote, or smoking, is done by a parent or other close relative, but in serious cases the specialist is consulted. Most illness is believed to stem from the intrusion by sorcery of a foreign object, which must be removed by suction, and shamans specialize in treating different kinds of sorcery. Some are able to restore a wandering soul to the body of the patient, who would otherwise face certain death. Payment is made whether or not the treatment is successful.

SOCIAL ORGANIZATION. Kayapó life is regulated not only by kinship obligations but also by several kinds of social institutions. Every individual passes through seven or eight successive classes during his lifetime, each one involving different kinds of privileges and new obligations to the community and to classmates. Between the ages of one and three, a child receives one of seven ritual names and becomes a member of one of four groups (two for boys and two for girls), placing him in a new kinship relation with other members of the same group. When a boy moves from the warrior class to the parent class, he joins one of the two men's associations, which function as cooperative and mutually complementary work groups in community projects.

The household group is a matrilocal extended family, normally composed of a woman, her married and unmarried daughters, and their children. The head and owner of the house is the oldest woman. Although a boy leaves his mother's house at about the age of ten to spend his nights and most of his days in the men's house, his closest ties remain throughout life with his sisters. It is a sister who paints a man for ceremonies, who cares for him in serious illness, and who prepares his body for burial. She also receives a choice part of any game he catches, and provides him with a refuge during marital difficulties.

A Kayapó village has one or two chiefs. If there is one, he is selected from among the leaders of the several men's associations on the basis of outstanding intelligence and courage, as well as seriousness and taciturnity. When there are two chiefs, they are the heads of the two men's associations. A chief exercises leadership because his experience and good judgment are respected, but he has no power to compel obedience. Aside from prestige, his position is rewarded by the gift of a piece of any large game brought in by a hunter and by possession of a house of his own. Order in the village is maintained by members of the warrior class, some of whom keep a vigil in the men's house during the day. Decisions involving the community, such as where and how large to make a new garden clearing, when to have a fish poisoning expedition, or when to set out on the summer march, are made in the evening when all males congregate in the men's house.

The men's house is the center of male social life. Although older heads of families spend increasing amounts of time during the day with their wives and children, they return every morning to greet the sunrise in the men's house. All of the male manufacturing activities take place here, and young boys receive their education in arts and crafts, tribal customs, and manly virtues under its roof.

LIFE CYCLE. From conception until about two months after birth, both parents observe dietary restrictions to insure that the pregnancy will be normal and that the infant will be strong and healthy. When the fetus begins to move, the husband goes to the men's house to sleep and does not return until the child is able to walk. Birth takes place in the company of close female relatives. Immediately afterward, the newborn is carried by a grandmother or paternal aunt from house to house to introduce it to the occupants. Its body is then painted with red and black designs, its hair is cut in bangs, and bands are placed around its legs. Three or four days later, its ears are pierced and, if a boy, also its lower lip. Until survival is assured, members of the family maintain a strictly reduced diet. Infanticide is practiced if twins are born or if the mother dies during or shortly after childbirth.

A child remains in the constant company of its mother for the first seven years. Nursing continues until age four or five, even if another child is born during this time. Parents rarely show anger to a child or reprimand it. After the age of three, its body is painted with elaborate black geometric designs, which must be touched up about every ten days to maintain their brilliance. Adoption is relatively frequent, not only for orphans but also for children whose parents remain alive. In the latter case, the child is given to a man or woman with whom one wishes to establish close bonds. The adoptive parent becomes a brother or sister of the real parent, and raises the child as his own.

From the age of about seven to ten, boys of different households play together, including practice with miniature bows and arrows. They may also accompany the men on hunting or fishing trips. Girls of that age begin to learn women's tasks, among them the art of body painting, and gradually assume an adult role. They will remain in their mother's house throughout their lives, in the company of their female relatives. Boys, however, are sent at about the age of ten to live in the men's house, where their education in tribal customs, arts and crafts, hunting, and warfare takes place. Each boy is assigned to a warrior, beside whom he sleeps and who serves as both instructor and model of the manly virtues of courage, endurance, and strength. Ordeals are employed to toughen him and inure him to pain. During this time his mother provides him with food, which he takes to the men's house to eat.

Following initiation at about the age of 15, a boy enters the warrior class and begins the most exciting period of his life. He enjoys the prestige of adulthood without family responsibilities; he occupies the best place at the center of the men's house; he is a featured performer in ceremonies; he devotes much of his time to keeping physically fit and ready for warfare; and he takes a leading role in maintaining village discipline and in training younger boys to become warriors in their turn. Young men hate to give up this privileged status, which they must do when their first child is born. Young women also are eager to avoid pregnancy, which they do by oral contraceptives and mechanical methods of abortion.

Marriages are arranged by the parents long before a child reaches puberty and sometimes shortly after birth. They normally take place when the girl is 10 to 12 and the boy 15 to 18. Monogamy is universal, but liaisons and adultery are frequent, although the latter may be ground for divorce if discovered. Initially, the husband continues to spend most of his time in the men's house, going to his wife's residence only in the evening. Until a child is born, the marriage bond is tenuous and easily broken by either party. A girl whose marriage fails may temporarily become the mistress of warriors.

After the birth of his first child, a man advances from the warrior class into the category of head of a family. As more children are born, he spends his time increasingly with his family, supervising the young children, obtaining food, and treating minor illnesses. Although he must be at the men's house each morning at daybreak, he ceases to feel embarrassed at being seen with his wife during the day. There is a strong rule requiring in-laws to avoid each other, and a Kayapó man never marries his wife's sister after the death of her husband.

Old men and women perform important functions in ceremonies and initiations, whether or not they are shamans. If a person becomes ill or infirm, he is cared for by a relative who will remain with him in the village during the dry season if he is unable to travel. When a man feels himself near death, he will seek out his sister's house. When he dies, his mother, sisters and widow lament loudly and cut themselves with knives, sometimes fatally. A widow shaves her head and does not remarry or participate in ceremonies or dances until her hair has regrown. The body is prepared for burial by female relatives, wrapped in a mat, and placed in the grave with various possessions. Poles and mats are laid over it for protection from the soil. Unless the proper ceremonies are performed, the soul may return to plague surviving relatives.

CEREMONIES. Kayapó ceremonies have two principal functions: the commemoration of mythical events of tribal significance, and the initiation of new members into one of the name or age groups. Before each festival, the men and boys go hunting for two or three weeks to obtain extra meat for the days ahead.

Among the special celebrations, which usually extend over several months, is one conducted by women in memory of a time when they were transformed into fish and the men were left alone and saddened. Another commemorates the gift of maize to the Kayapó and helps to insure the maturation of the new crop. Some rituals, such as the purification of a victorious warrior or the initiation of boys into the warrior class, are participated in only by the individuals directly involved until the finale, in which the whole village takes part. Members of other communities are never invited.

Although the specific ceremonies, dances, and costumes vary with each occasion, most festivals end in the same way. When possible, the conclusion is timed to coincide with the full moon. Everyone puts on all their ornaments and joins a solemn procession at sunset. The only ones excused are the ill and the old women, who grieve for the past that is gone beyond recall, for those who have died, and for their own lost youth. Singing and dancing continue through the night under the light of the moon.

TRADE. No accounts mention the existence of trade during the aboriginal period. Today, Brazil nuts are exchanged for knives and other kinds of European goods.

WARFARE. A Kayapó man achieves full adult status only after having killed a person in warfare. Consequently, the Kayapó have earned a reputation for belligerence, which has been expressed in recent times by attacks on Brazil nut and rubber gatherers. Members of the warrior class maintain themselves in a state of readiness for attack or defense, and keep watch throughout each night. Anyone not a member of the village is a potential threat, and hostility is more intense toward related groups than toward alien tribes.

The chiefs of the two principal men's associations are the war leaders, in whose hands the strategy rests. Before an expedition, they harangue their followers to bolster their courage and arouse enthusiasm. The weapon of warfare is the club. It is not necessary for each person to make a kill, and all participants in the expedition can share the glory of victory by striking an enemy, even if the person is already dead. The defeated village is burned, and women and children who have not fled or been killed are taken prisoner and adopted into the conquering tribe. A warrior who has killed an enemy must submit to purification rites, which involve dietary restrictions, confinement to the men's house, and chest scarification in the form of a large V extending from shoulders to navel.

Religion and Magic. The Kayapó conceive the world as filled with evil spirits dedicated to the persecution of human beings. These invisible enemies fall into four categories: the souls of the dead, a kind of inanimate power projected at will by people or animals, a water monster, and a mythical bird. All may cause illness or death, and the principal means of protection is to remain with the group. A Kayapó consequently avoids solitude (which is not a difficult thing to do in the context of village life). The funeral ceremony is designed to please the soul of the deceased, so as to reduce its sorrow at leaving the land of the living and its animosity toward those who still enjoy life. Both red paint and tobacco smoke afford protection against spirits, and since women are particularly vulnerable to attack, it is they rather than the men who do the most smoking. Only a shaman is able to establish an amicable relationship with a spirit, whose help makes it possible for him to extract objects that have been introduced into the body of a victim by sorcery.

THE SIRIONO

Location and Environment. On the southwest margin of the tropical forest, between the Brazilian highlands and the foothills of the Andes, the elevation is only 500 feet. Annual rainfall is about 80 inches and is concentrated between October and May. This combination of low land and heavy precipitation produces extensive inundation during half the year. Even during the dry season, rivers and lakes are numerous. The present habitat of the Sirionó, south of the Rio Guaporé between 63° and 65° west longitude (Fig. 6), is dense forest. A little to the west, where the Sirionó formerly lived, the forest shrinks to islands in a grassy plain. The typical Amazonian flora and fauna prevail.

The Sirionó, speakers of a Tupian language, have gradually declined with the encroachment of civilization. Their population was estimated as less than 1,000 in 1825. Although some have been exposed to missionary influence or have settled on farms and cattle ranches, many still live under aboriginal conditions. Census figures on two bands in 1940 gave a sex ratio of one male to about 1.2 females.

SETTLEMENT PATTERN. The Sirionó live in bands containing between 50 and 100 individuals, which camp in one spot for several months during the rainy season but move every five or ten days during the rest of the year as game in the vicinity is exhausted. Bands tend to be spaced about eight to ten days of travel apart; although they do not occupy clearly defined territories, they try to avoid each other. If members of one band run across another's hunting trail, they will retreat in the opposite direction. Travel is by foot, since the Sirionó possess no watercraft.

The rainy season campsite is a place above flood level where there is little underbrush to clear away. A house about 80 feet long and 25 feet wide, capable of sheltering 60 to 80 persons, can be erected in an hour. All the men share the work, each family constructing the portion it will occupy. Several large trees are selected to serve as uprights, against which a frame of poles is lashed. The wall consists of palm leaves, almost ten feet long, stood on end against this frame to lean inward at an angle of about 60 degrees. Variation in stem length results in incomplete coverage at the ridge, leaving gaps through which smoke escapes and rain enters. On rainy nights the inhabitants must leave their hammocks to squat by the fire (which is protected) if they wish to remain relatively dry. There are no doors, and entry is made anywhere through the wall. A similar but even more flimsy structure is erected for overnight camps.

The interior is crowded with hammocks, which are hung about

three feet apart across the width of the house. Each adult has his own hammock and children sleep with their mother. The chief's family occupies the center. A fire between every two hammocks serves for cooking and heating. Lightweight possessions are stored in baskets and gourds hung from the palm midribs; pottery vessels are left on the floor.

Dress and Ornament. Neither men nor women wear any clothing. A necklace of chonta palm seeds, feather quills, and coati or spider monkey canines is the only common ornament. The face (and sometimes the body) is rubbed with red paint, and feathers are glued to the hair on festival occasions. Both sexes wear their hair cut short.

Subsistence. The Sirionó are principally hunters and gatherers, relying only secondarily on agriculture and fishing (Fig. 14). Most gardening is done at the beginning of the rainy season, although planting may occur at any time of the year. Husband and wife work together to clear a patch about 45 feet square near the house, where they plant maize, sweet manioc, and sweet potatoes. Cotton, tobacco, calabashes, and achiote are typical non-food crops. The only agricultural tool is the digging stick. If a man has a favorite hunting spot, he may make another garden there to exploit when he is in the vicinity. There is no weeding, nor is any kind of magic practiced to promote the growth of the plants.

The most important male activity is hunting, and a man spends at least 50 percent of his time in this way. He leaves with his bow and arrows before dawn, either alone or with a relative. Although he often travels 40 miles and is adept at stalking and decoying birds and animals, he may return empty-handed. Usually, however, he will shoot a monkey or a bird. The Sirionó possess no dogs, making game such as tapir and peccary more difficult to obtain than would otherwise be the case. No animal is considered inedible, with the exception of snakes. A hunter should not eat the meat of an animal he has killed, however, or the species will not return to be hunted. Various kinds of magic are employed to attract game.

Fishing also is done with the bow and arrow. It is relied upon mostly in the dry season when the fish concentrate in the shrinking lakes and streams and the water is more transparent so that the target is more easily visible. Only four of the numerous species potentially available are usually caught. A common tech-

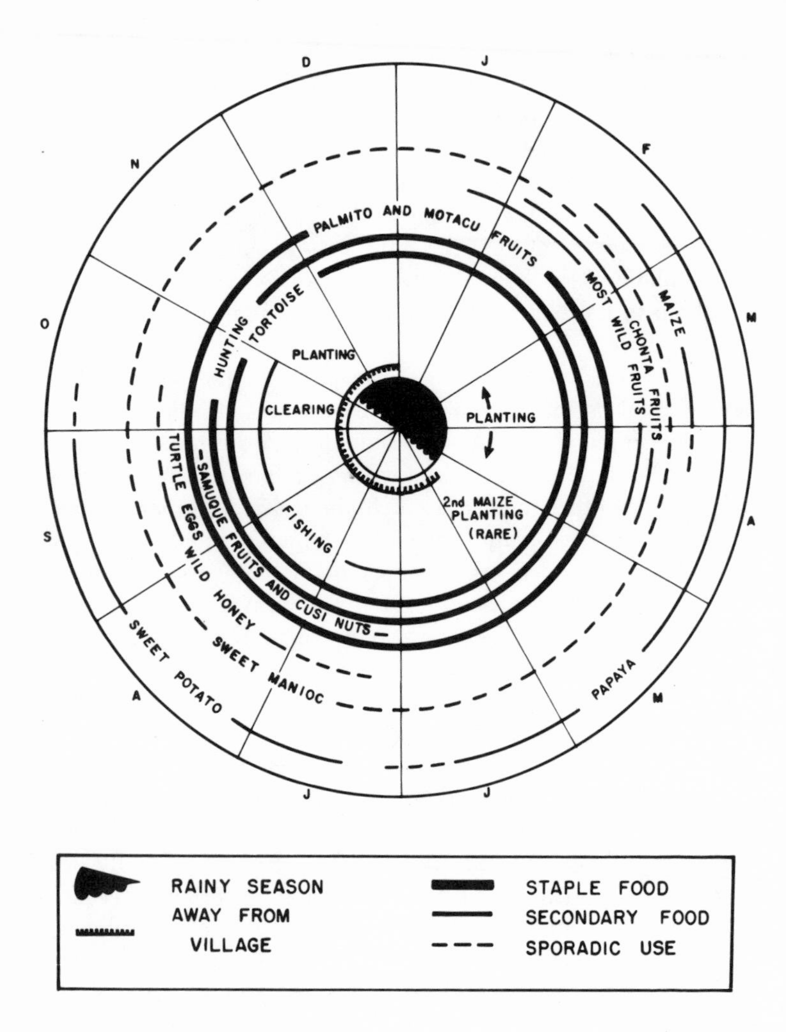

Fig. 14. Annual subsistence round of the Sirionó. Game, land tortoises, palmito, and motacó fruits are the primary year-round staples. The principal agricultural crops—maize, sweet potatoes, and papaya—are available only for a few months. Sweet manioc is sporadically consumed throughout the year. A great variety of fruits, nuts, and other kinds of wild foods is eaten during most of the year. The marked seasonality of rainfall has led to a pattern of alteration among the Sirionó between dry season wandering and rainy season sedentariness.

nique is to drop fruits into the water and shoot the fish as it surfaces to eat the bait. A man may obtain up to a dozen fish a day in this way.

Gathering, which is second in importance to hunting, is usually done in nuclear or extended family groups. Women and children spend more time in this activity, but if trees must be climbed, this is a man's job. When seasonal fruits are harvested, as many as possible are eaten on the spot before baskets are filled to be taken back to the camp. Palmito is the principal staple because of its constant availability. Tortoises weighing up to nine pounds are relatively abundant and may be captured alive to be consumed at a later time. No insects and few roots are eaten.

People frequently complain of hunger, but actual starvation is not a serious threat. Most food is eaten the day it is obtained, with the result that intake is very uneven. When the occasion arises, four adults can dispose of a 60-pound peccary at one sitting; more frequently, however, the group passes two or three days without eating meat. No effort is made to maintain a steady food supply, and food is not sought until everything on hand has been consumed. Many fruits are eaten raw; meat and garden produce are boiled or roasted. Maize is roasted on the cob or ground up to be boiled with meat or made into cakes. A mildly intoxicating drink is produced from cooked sweet potato, manioc or corn meal, water, and honey, which are mixed together and allowed to stand for three days. This is consumed principally during the dry season.

OTHER ACTIVITIES. Although certain tasks are customarily performed by males and others by females, it is often necessary for one sex to engage in an activity traditionally allotted to the other. For example, a man may weave a basket to carry home food, although basketmaking is considered women's work. In addition to building shelters, men make wooden objects, such as digging sticks, spindles, bows and arrows, mortars and pestles, as well as containers and utensils of calabash. Women spin cotton, weave hammocks, baskets, and mats, make cord (including bowstring), necklaces, feather ornaments, pottery vessels, and pipes. Household activities, such as carrying water, fetching firewood, cooking, drying tobacco, and caring for children, also are women's work. Although all adults carry burdens, men do not transport women's things when moving camp. There are no specialized occupations of any kind; even shamans are unknown.

Social Organization. Each band is an independent social, political, and economic unit. It is composed of four or five matrilineal extended families, each in turn made up of three or four nuclear families. Marriage normally occurs within the band. Polygamy is restricted to the band chief and one or two other outstanding hunters, and plural wives are usually sisters. The average number of children per family is two.

The band is most cohesive during the rainy season, when it remains sedentary. Except for house building, however, there are no social or economic activities that involve the entire community. The largest permanent cooperating unit is the extended family. Male members often hunt together, while the women go off as a group to collect wild fruits. When more food is obtained than can be consumed by a nuclear family, it will be distributed within this larger group. In spite of reciprocal kinship obligations, however, sharing is often done unwillingly and food may be consumed in secret to evade demands by relatives.

The band chief has no authority, and although he may suggest a hunting trip or the direction that the band should move, there is no obligation to obey or punishment for not doing so. He does not arbitrate in disputes or officiate in any ceremony. He maintains his prestige by being the best hunter, the best song composer, and the best drinker, and is rewarded by possession of several wives and the allocation of the central part of the house to his family. He will be succeeded by a son or younger brother, the choice falling on the outstanding hunter. Disputes are settled between the persons involved, and everyone is expected to defend his own rights as well as to fulfill his obligations. Quarrels frequently break out between members of a family, most of them over food distribution. If the dispute involves members of different extended families, relatives will intervene to bring it to an end. Except in drinking bouts, arguments rarely lead to violence. If hostility is too intense, one party will either join another band or split off with relatives to found a new band. Murder is very rare and results in banishment of the murderer for an indefinite period of time. Sorcery is never employed for revenge.

Life Cycle. When a wife becomes pregnant, both prospective parents observe various dietary restrictions in order to prevent the infant from being physically deformed or developing undesirable psychological traits. Birth takes place in the house before an audience of women and children, none of whom attempt to assist the mother. If the event occurs during the day, the father leaves at the onset of labor pains to go hunting, since the child will be named for the first animal killed. Infanticide is not practiced, even if the infant is born with a clubfoot, as is frequently the case. Although boys are preferred, girls and deformed children are treated with equal affection.

Both parents restrict their activities for three days following birth, remaining close to their hammocks, observing dietary taboos, and scarifying their legs. These procedures are intended to ensure the health of the child. On the day after birth, the infant's eyebrows and forehead hair are removed and made into a necklace, which is worn by the mother. On the third or fourth day special rites terminate the restrictions on parental activities, except that sexual relations are not resumed for about a month.

Young children are treated with great affection by both parents, and except for the painful ordeal of depilation of eyebrows and forehead about every two weeks, the infant is indulged whenever it begins to cry. Although weaning does not take place before about three years of age, premasticated food is provided after about six months. About the time of weaning, a boy receives his first bow and a girl a miniature spindle. By the age of eight a child has learned most adult tasks, and boys begin to go hunting with their fathers to perfect their tracking ability and knowledge of the habits of game. By this time, a child has learned the food taboos necessary to avoid illness and is expected to take responsibility for observing them. Expressions of aggression in young children are encouraged so that they will be prepared to defend their rights in adult life.

There is no puberty observance for boys, but girls must undergo a brief ceremony before they become eligible for marriage. Whenever several girls are approaching puberty, their heads are shaved by their parents and they are taken to a spot near water. A platform is erected, on which the girls sit for two or three days. During this time they are bathed frequently and are instructed in the food restrictions that they must observe until they are married. On their return to the village, the girls prepare themselves for marriage by working hard at all kinds of household tasks. They must refrain from intercourse until their hair has reached shoulder length, which takes about a year.

The preferred marriage is between a man and his mother's brother's daughter, and about 50 percent of the unions are of this type. The remainder are between classificatory cross-cousins. Brothers often marry sisters, and a man's second wife is likely to be his first wife's sister. There is no ceremony or property settlement, and the marriage begins when the husband moves his hammock from the area occupied by his family to a place next to his wife. Divorce is easy but uncommon. Adultery is not punished unless it becomes too flagrant, in which case the wife will be disowned and will immediately remarry. The principal way for a man to achieve status is by being a good hunter and by having several wives and many children. A woman gains prestige by bearing numerous children, by her skill as a provider, and by being married to a good hunter.

The pace of life is rapid for a Sirionó. He begins to contribute to the food quest while still a young child, is married by puberty if not before, and begins to decline in strength and prowess after 30. By 35 he has reached old age and has begun to consume more food than he can provide. If too weak or ill to keep up with the band, he will be left behind to die. Illness is believed to be caused by evil spirits, who enter through the nose or mouth during sleep. Since there is no treatment, it occasions great anxiety on the part of the affected individual, who takes to his hammock until cured or dead. Under these circumstances, few live more than 35 years, and it is consequently a rare individual who sees the birth of a grandchild.

When death approaches, the dying person is removed from his hammock and laid on a mat, around which weeping female relatives sit to await the end. After death occurs, the body is wrapped in two mats and placed on a platform in the house along with gourds of water, pipes, and a fire, but no food. The men shoot arrows through the walls in all directions to drive out spirits, and the house is abandoned. Close relatives continue formal mourning for three days, during which they fast, scarify their legs, and decorate themselves with feathers. About a week later, the surviving spouse remarries, a widow usually marrying her husband's brother. After the flesh is gone, the relatives must return to bury the bones in order to prevent the return of the soul in the form of a monster which may bring illness to the family. The skull is kept, however, for protection against disease and death.

CEREMONIES. The only public ceremony performed by the Sirionó is that marking entry into adult life following the birth of the first child. It is held annually during the dry season, when the wild honey needed for preparation of an intoxicating drink is abundant. While the drink is fermenting, the participants have their hair cut and are painted with red achiote and ornamented with feathers. Drinking takes place in two groups, one composed of men and the other of women, and is interspersed with singing and dancing. After they are partly drunk, male participants pierce each other's arms and those of the women with a stingray spine, allowing the blood to drip into small holes in the ground. The next morning the men go hunting and the women gather palmito, returning to consume the remaining drink during the afternoon. New pots must be used for cooking and special food taboos observed to prevent the wounds from infection. This ceremony is believed to rejuvenate the participants and also to attract game.

TRADE. No articles are exchanged within or between bands.

WARFARE. The Sirionó make every effort to avoid hostilities with other bands or neighboring tribes.

RELIGION AND MAGIC. The Sirionó believe that the vicissitudes of life, including accident, disease, bad luck, cold weather, and death, are caused by invisible evil spirits. They cannot be controlled or pacified, and the only way to evade them is by obeying tribal customs. Another category of supernatural enemy takes the form of a large, ugly monster, which lies in wait during the night to carry people off and strangle them. Ghosts of the dead are likely to return in this form if funerary rites are not properly performed.

Since the supernatural world is viewed as immune to human influence, there are no individual or group rituals. Dietary restrictions, magical procedures, singing, and dancing are believed to increase game, however. Dreams, which occur when the soul wanders, foretell the future.

THE WAIWAI

LOCATION AND ENVIRONMENT. The Guiana coast is separated from the Amazon basin by a low range of mountains that now coincide with the international boundaries between Brazil and the countries to the north. The habitat of the Waiwai is part of the geologically ancient Guayana highlands, drained by the headwaters of the Essequibo, which flows north to the Atlantic, and the Mapuera, which flows south into the Amazon. The terrain is undulating, and during the rainy season forms an irregular patchwork of high spots interspersed with inundated areas. Precipitation is heavier from April through July and December to January, but rain can be expected any day of the year. Brazil nut trees, large river turtles, and caymans are rare north of the divide, otherwise typical Amazonian flora and fauna prevail.

This watershed area is occupied by several tribes of Carib, Arawak, and independent linguistic affiliation but generally similar culture. All are relicts, having been decimated by European diseases. The Carib-speaking Waiwai were reported in 1837 to have a population of about 150, distributed in three villages south of the Acarai mountains. In 1955, they numbered 170 and were living in seven villages distributed on both sides of the mountains (Fig. 6). In the interim, however, they were almost exterminated (whether by enemy tribes or disease or both is not clear), and the present Waiwai tribe has been reconstituted by the assimilation of individuals from neighboring tribes, including the Arawakan-speaking Mouyenna, the Taruma, and the Cariban-speaking Parukoto. By 1958, all the Waiwai villages were located on the upper Essequibo. Rapid deculturation took place between 1955–1958, but prior to that time European influence was limited to the acquisition of glass beads, knives, hoes, metal containers, and a few guns, and to medical aid. The recent population is unbalanced sexually, with males outnumbering females about three to two.

Settlement Pattern. Each Waiwai village consists of a single circular communal house, which varies in size according to the number of occupants. It is located close to a creek and within easy access to the river, where there is sufficient high land for gardens to be raised for several years. Because of the irregularity of the terrain, there is seldom room for more than three fields, and consequently villages are moved about every five years. If the house deteriorates before the fields are exhausted, another is constructed nearby. It is also abandoned and burned on the death of the head man, the shaman, or the wife of either man (and formerly, abandonment occurred on the death of any adult resident).

The house is constructed by the men who will occupy it, while the women prepare the beverages to be consumed in the inaugural feast. A ring of posts forming the wall is supplemented with a smaller interior ring that helps to brace the roof when the center pole is removed, after construction is completed (Fig. 15). The vertical wall is about five feet high, while the roof rises to some 30 feet in the largest houses; the diameter ranges from 35 to 65 feet. Thatch covers the exterior so that the only illumination aside from the fires is provided by two doors at opposite sides. Most work consequently takes place in shelters constructed adjacent to the house. Platforms are erected around the outside on which to dry cassava bread, which may alternatively be spread on the lower edge of the house roof. The small clearing is kept swept. Several paths lead in different directions, to the river, to the gardens, or to a nearby water source.

There are no partitions inside the house, but each nuclear family has a defined area. The village head takes a favored spot at one side, away from the doors, but locations within the house otherwise are optional. The hammocks of a nuclear family are hung to form a triangle between an inner post and the wall, surrounding a hearth. The husband's hammock is suspended above that of his wife, who keeps a fire going during the night for warmth. After infancy, boys sleep with their father and girls with their mother until about age eight, when each is given a separate hammock. An unmarried brother or sister or an elderly parent of the wife may complete the family group. Along the wall behind the hammocks are the platforms where the family dogs are kept. Baskets, gourds containing hair oil or paint, feather ornaments, musical instruments, and other possessions of the men are hung on the wall or from the rafters; bows and arrows are stuck into the thatch. Female utensils, such as cooking pots, gourds, fire fans, and mats, surround the fireplace or are placed on a rack suspended from the rafters. A communal fireplace occupies the center of the house floor.

DRESS AND ORNAMENT. A single article of clothing is worn by both sexes. For the men, this is a narrow cotton loincloth; for the women it is an apron woven of glass (formerly, seed) beads.

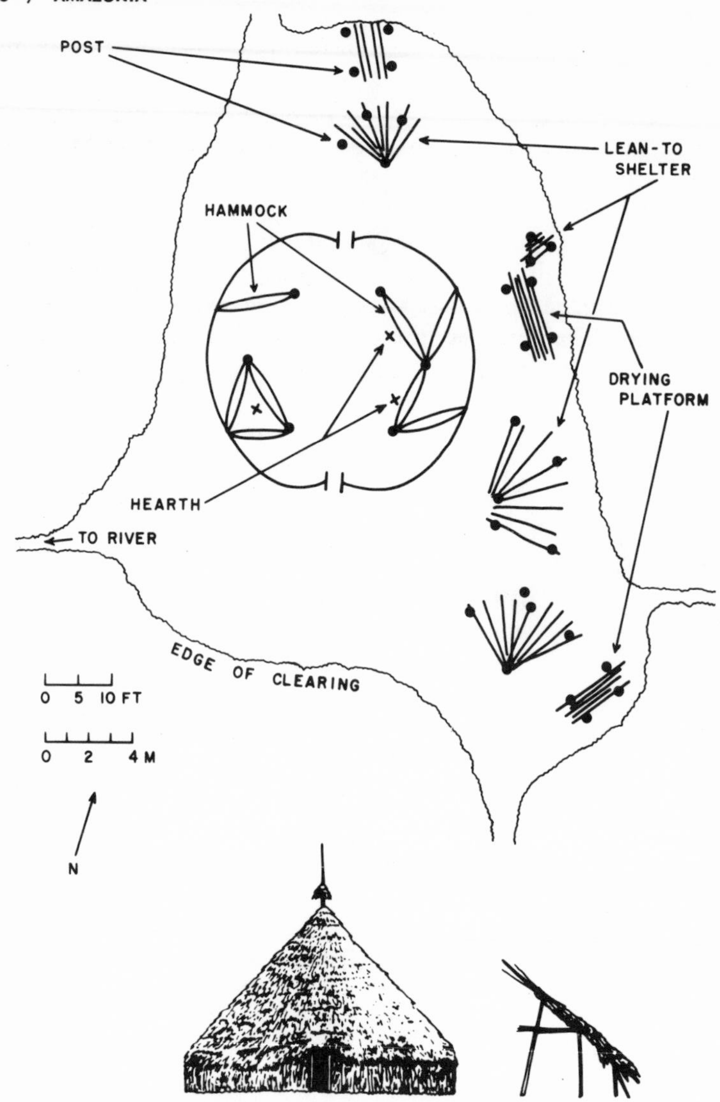

Fig. 15. Ground plan and side view of a Waiwai communal house occupied by a matrilocal extended family. Hammocks are suspended between roof supports and the wall, each segment being occupied by a nuclear family. Fires (X) are kept burning at night for warmth. The clearing around the house contains fan-shaped shelters, where craft and household activities are carried out, and racks for drying cassava bread. Paths lead from clearing to riverbank, gardens, and forest.

Adults wear a long string of white beads wound around the upper arm, a similar string of blue beads below the knee, and conical ear ornaments faced with a shell disk. Both sexes cut their front hair in bangs; the women gather the remainder into a bun, while the men bind it into a long tail which is sheathed in a cane tube for daily wear. A variety of necklaces, bracelets, and other small ornaments as well as body and face painting are utilized according to the taste of the individual. Festive occasions call for more elaborate feather crowns, nose ornaments, belts, and streamers, which are worn principally or exclusively by men.

SUBSISTENCE. The staple of the Waiwai diet is bitter manioc, supplemented by yams, sweet potatoes, squash, and fruits, such as pineapple and papaya (Fig. 16). Bananas, plantains, and sugar cane have been introduced in post-European times. Gourds, cotton, tobacco, silk grass, and achiote are also grown. Clearing of the forest takes place from August to September under the direction of the village chief. All men and boys participate, except a few who hunt for the community. Small trees are cut first and then large ones are notched so that felling one tree at the edge of the field will knock down others in its path. After drying for about six weeks, the slash is burned and the field is divided into family plots. Planting is timed with the December rains and is done by men, who may be assisted by their wives. Small hills are made for manioc, yam, and squash, but there is no other cultivation. Manioc is planted last. Occasional selective weeding is done during the first two years. Manioc cuttings are replanted as the roots are harvested until the third year, when the plot is abandoned.

Manioc processing consumes a large part of a woman's time. Although several women may work together, they all do the same tasks. The roots are harvested early in the morning and peeled immediately upon return to the house. Grating is done on a wooden board imbedded with small stone teeth, and the pulp is accumulated on mats or in pots or an abandoned dugout canoe. Four women can peel and grate about 120 pounds of tubers in seven hours. The next step is pressing the pulp in a square sifter with a little water, which removes some of the poisoned juice. The remainder is extracted by squeezing in a flexible tubular basket, known as a tipiti. The resulting cylindrical lumps are dried on a rack hung above the fireplace, and then broken into

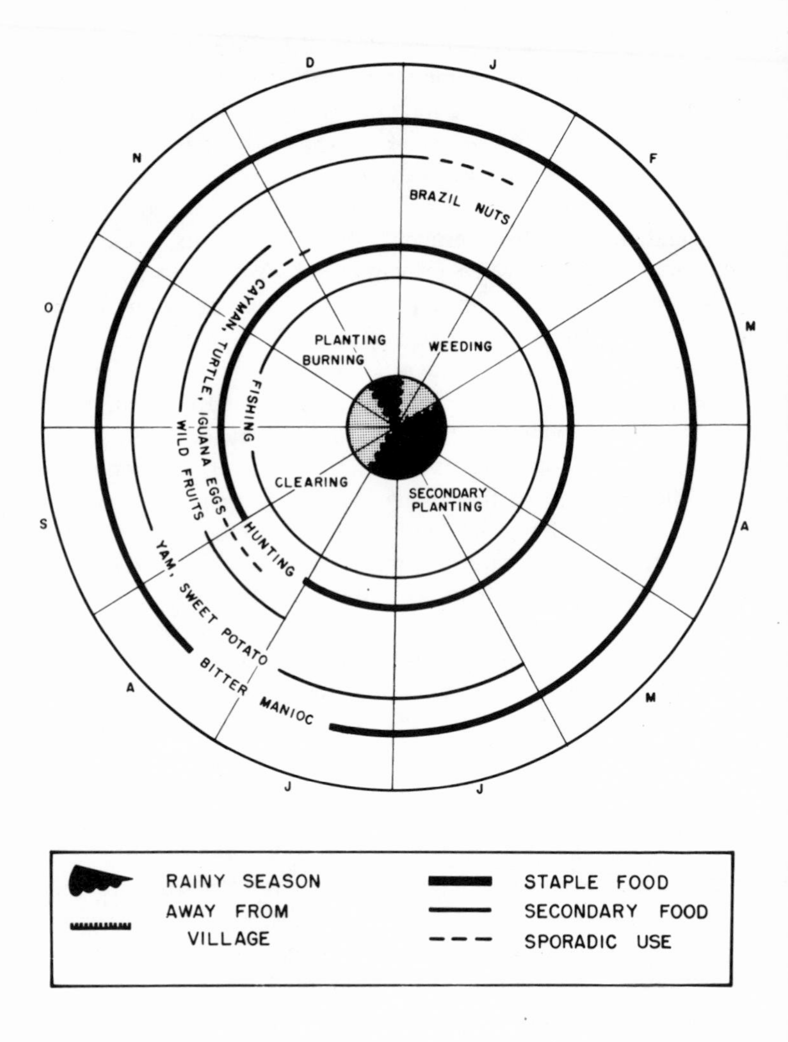

Fig. 16. Annual subsistence round of the Waiwai. Game and bitter manioc are the year-round staples, supplemented by fish. Yams and sweet potatoes are secondary crops. Although wild fruits, turtles, eggs, and other foods are exploited during certain seasons, gathering is not an important subsistence activity. There is no well-defined dry season and the group remains sedentary throughout the year.

pieces small enough to be pulverized in a wooden mortar. After the flour has been sifted, it may be eaten alone or mixed with water, or baked with a variety of embellishments.

The principal product is cassava bread, which is made by spreading a layer of meal about ⅜-inch thick on a large circular griddle over the fire. When the lower surface is light brown, the wafer is turned over. If subsequently dried in the sun, the "bread" will keep for some time. It is usually torn into pieces and dunked in pepper pot or water, or wrapped around a piece of fish or meat. When dipped in water and laid aside to ferment for several days, it forms the basis for several beverages. The poisonous juice is also utilized. Some fourteen varieties of bread and at least thirteen beverages are prepared with manioc flour or juice as the principal ingredient. Drinks are varied by the addition of crushed bananas, gourd seed, or Brazil nut.

Hunting is the favorite pastime of the men, who go out with their bow and arrows before dawn in the company of a relative. Birds may be decoyed by imitating their calls, while tapir, peccary, and deer are usually hunted with dogs. All animals and birds are eaten except carnivores and the opossum. Not only the meat but the liver, heart, brain, and intestines are often consumed. Being a good hunter is a way to achieve prestige, and a variety of magical practices is employed to enhance the possibility of success. These include dietary restrictions, scratching the skin with squirrel claw, exposure to biting ants, and special face painting. Lizards, frogs, iguanas, and turtles are eaten, and the latter may be captured alive and saved until needed.

Waiwai men also apply their skilled marksmanship in fishing, shooting fish from a canoe, the shore, or a scaffold erected in a shallow part of the river. Traps are also set, and occasionally (especially when ponds are isolated by the falling water), poisoning is undertaken with the help of women and children. Most species are eaten, and if the catch is larger than can be consumed the excess is preserved by smoking.

Gathering is a minor source of food, but palm fruits, cashew fruit and nuts, Brazil nuts (also processed for hair oil), honey, wasp eggs, and the eggs of iguana, turtle, and cayman are all eaten in season. Plants supply the raw materials not only for tools, weapons, and ornaments, but also for fish bait, arrow poison, black paint, medicines, and many types of containers. Men do most of the gathering, often when returning from fishing and hunting trips.

Other Activities. In addition to building houses and clearing fields, men produce many articles for personal or family use. They make canoes, bows and arrows, and musical instruments (drums, flutes, panpipes, rattles); carve stools, paddles, grater boards, and gourd containers; twist ité palm cord and silk grass string; weave hammocks, cotton loincloths, and all kinds of baskets and mats. They also make feather ornaments and dance costumes. Women carve spindles and spin cotton, make bead ornaments and aprons, insert the teeth in grater boards, prepare achiote paint, and make pottery. In addition, they have daily household tasks, such as carrying water, collecting firewood, tending the fires, and preparing food and drinks. Both men and women care for young children, paddle canoes, and carry burdens.

The principal weapon is the bow and arrow. Bows are 6.5 to nine feet long, and arrows are of comparable length. The arrow has a reed shaft, a hardwood foreshaft, and a point of bamboo, wood, bone, or metal. A removable harpoon point is employed for some fish, peccary, turtles, and cayman. Points intended for monkeys are poisoned.

Although all adults know how to make everything assigned to their sex, all are not equally skilled. Consequently, a woman who is an unusually good potter or a man who weaves particularly attractive baskets may exchange his products for something else. The only true specialized occupation is shamanism. A young man who exhibits special aptitude or is "called" by a dream goes for instruction to an older practitioner. In addition to effecting cures with the aid of tobacco, red pepper, stinging ants, and magic, he is able to interpret dreams and cause illness or death. His chief reward is prestige, since he is not relieved of normal family responsibilities.

Social Organization. The Waiwai tribe is a linguistic and social unit, but the largest economic and political entity is the village. The occupants are families related through the female line (for example, a woman, her married daughters and their families; or several sisters and their husbands, children, and matrilineal relatives). The house leader, or village chief, is a man who commands the respect of the others and who is able to carry out the responsibilities of the office, which include receiving and feeding guests, providing for old people who have no young female relatives, issuing invitations for feasts, knowing how to make a good field, predicting rain, and organizing community projects, such as fish poisoning and field clearing. He is usually also a shaman, although this is not required. Because of his hospitality obligations, the chief must have a wife, and if his wife dies he will be deposed if he does not remarry. When a village moves, some members may split off to found a new community under a new chief. Except for prestige and the right to occupy the best location inside the house, the only special recognition accorded a chief is the burning of the house at his death or the death of his wife.

Interpersonal relations are regulated by kinship, and each person knows his rights and responsibilities with reference to those whom he calls by various kinship terms. If a conflict develops between two individuals, either because the request for an article is refused or a service is not rendered, the injured party may ask his adversary to engage in a formal dialogue. Thereafter, if the latter continues to refuse, he runs the risk of blood vengeance. This ritual dialogue is also employed to arrange marriages, extend an invitation to a feast, deny an accusation of sorcery, and on many other occasions. It provides an acceptable mechanism for reaching agreement without involving a third person as an arbiter or engaging the community in conflict, which everyone dislikes.

In general, the Waiwai place great value on independence. The existence of reciprocal kinship obligations increasingly restricts an individual's freedom, however, as village size increases. Also, the larger the population, the greater the danger of sorcery. In this context, a strong desire for independence becomes an important factor in keeping Waiwai villages both small and widely separated.

LIFE CYCLE. During the last three or four months of a woman's pregnancy, both parents must refrain from certain activities and observe dietary restrictions. A woman must not make pottery or participate in fish poisoning expeditions. Birth occurs in a special hut, with the assistance of the husband or a female relative. Infanticide may be practiced if twins are born, or if four previous children are of the same sex as the newborn. Both parents remain in the birth hut for two weeks, and are subject to various taboos during the first three years of the child's life, while his soul is not yet permanently attached to his body and consequently is in danger of being lost. The father should not hunt or fish except for small fish, and the mother should not eat meat during this period. Although these restrictions are not strictly observed, the limitation on male activity makes the father available to watch the baby while the mother is occupied with manioc preparation. Children are kindly treated, especially by their grandparents. Weaning takes place about the age of three if another child has been born. About the age of five, girls begin to help their mothers dig manioc roots, while boys begin to practice marksmanship with miniature bows and arrows.

Boys pass puberty with no special recognition, except that between the ages of 13 and 15 they receive an upper arm band and have their septum pierced—both badges of adult status. At the onset of menstruation, girls are isolated in a special hut with no doors. They remain secluded for about two months, speaking to no one, subsisting on cassava flour and juice and small fish, and spinning cotton to pass the time. On release, they receive upper arm bands and may henceforth have sexual relations with potential husbands. For the next two years, they must continue to observe many food taboos and are kept working hard.

A girl is marriageable after initiation. Marriage is arranged during a formal dialogue between the fathers of the groom and the bride, and usually involves payment of a bow or a hammock to the girl's family. After agreement is reached, the new husband hangs his hammock above that of his wife and married life begins without further ceremony. No payment is made for a second wife. Husband and wife enjoy relatively equal status in marriage, and adultery was formerly punished by death. Polygyny is permitted and co-wives are preferably sisters. Polyandry also occurs, but is not common. For a variety of reasons, among them marked age difference at first marriage, death, incompatibility, and laziness (but not sterility), most adults have a series of spouses. Because of the shortage of Waiwai women, several men have married women from other tribes; children of such marriages are considered Waiwai.

During their adult life, Waiwai men and women pass through three age classes, each of which is associated with different terms of address and reference. The first is entered at puberty and ends for girls with the birth of the first child and for boys with marriage. The second extends until about the age of 45, and is the most active period of life. The third is the time of old age, to be spent relaxing with grandchildren and engaged in lighter activities, such as basket weaving for men or spinning for women. Elderly persons are treated with respect but suffer loss of prestige, particularly if the death of a spouse leads to dependence on a younger relative or the head of the village.

Except in the very young and very old, death is believed to be caused by sorcery. A dying person is moved to a shelter near the house, where the spouse or a relative stays with him until the end. The body is cremated the same day or the next morning and all personal property is destroyed except for a man's ax or a woman's bead apron, which passes to a son or daughter. Adult female relatives and children cut off their hair, while male relatives trim the ends of their pigtails. When the cremation fire has died down, the nearest male relative collects unburned bones to use in the magic "blowing" that will avenge the death. He often will not know who the guilty person is, but if the sorcery is successful the culprit will die within two months.

CEREMONIES. Two principal dance festivals are held at irregular intervals for reasons of a secular nature, such as the inauguration of a new house, the reciprocation of hospitality, or simply to provide an opportunity for social contact, especially between marriageable individuals. Each village sponsors a dance at least once a year and invites the members of one or more other villages, the number depending partly on the available food supply. The most common dance is the shodewika, which lasts three to seven days, and in which both men and women participate. The yamo dance, held less often, lasts for about two months with lapses for essential subsistence activities. During the first month, men dance to flute accompaniment out of sight of women; during the second month, they dance to rattles and women join in the singing. Neither dance is said to have any ritual significance (although certain features suggest this once may have been the case). A dance formerly performed in early January after planting has been abandoned, and no other feasts or village ceremonies have been reported in connection with subsistence activities, life crises, or other similar events.

TRADE. A few essential articles incorporate raw materials not available in the Waiwai territory and must therefore be obtained by trade. The most important are arrow reed and shell disks for ear ornaments. The latter come from the lower Mapuera, inhabited by the Shereó; the former are obtained from the Mouyenna and Wapishiana. Bows and certain other articles available locally may be acquired by trade because of the opportunity this provides for social interaction. European goods, such as glass beads, axes, knives, and hoes, are secured from the Wapishiana in exchange for grater boards and hunting dogs. Trading is an individual activity and no one has a right to dispose of another's property—even that of a small child. On the other hand, it is considered bad form to deny a request.

WARFARE. The Waiwai have no tradition of warfare and do not engage in raids. At the present time they have no close neighbors to threaten their own tranquility.

RELIGION AND MAGIC. Human beings and animals (and some plants) have souls or spirits with habits that may be dangerous. The human soul is not permanently attached to the body during its first three years, so there is a risk that death may result from soul loss. The parents must take many precautions to prevent this. Certain animal spirits may cause illness if they see a person, but since the spirits are blind to red, the Waiwai paint their bodies with achiote as protection. Bush spirits will kill people who wander in the forest alone at night. Only the shaman is able to command certain spirits, which he does to effect cures.

The principal technique of manipulating the supernatural world is magic blowing. If one wishes to kill another person, the most effective method is to blow over his sleeping body. This is dangerous, however, not only because of the possibility of discovery but also because magic wind might reverse the air back to the blower with fatal consequences. Alternatively, one may obtain something recently in contact with the victim, such as food, nail or hair clippings, or a personal possession (all of which are carefully hidden to forestall such use), and blow over it. When a person becomes ill or feels a pain, he knows that magic blowing has been practiced on him. Unless he can discover the culprit he is likely to die. Although the most common motive for magic blowing is revenge for the death of a relative, it may be practiced against a thief or someone against whom one has a grudge. Magic blowing may also be used to cure sickness or to bring success in hunting or a bountiful harvest.

Chapter 3

ADAPTIVE ASPECTS OF TERRA FIRME CULTURE

GENERAL CONSIDERATIONS

The five tribes selected as examples of adaptation to the terra firme environment exhibit numerous cultural similarities. They subsist on many of the same plants and animals, which are obtained by similar methods; they live in extended family groups in communal houses; their life cycle begins with an indulgent childhood, proceeds into adulthood via initiation rites and an arranged marriage to similar kinds of community and family responsibilities, and ends at a relatively young age in infirmity and death. The invisible world is usually pictured as hostile, and sorcery is greatly feared. Clothing is minimal, but ornaments are frequently profuse and colorful. The division of labor is along sex lines, with hunting and fishing always the responsibility of men. Although there is a chief, his authority to command is minimal and his position does not relieve him of routine male duties. The only real specialist is the shaman, who is able to communicate with supernatural beings or forces. Periodic festivals serve more to promote social solidarity than to placate the supernatural.

These general similarities have led to the recognition of a tropical forest culture area, with boundaries generally coinciding with those of the natural geographical region. One reason for this correlation is the presence of the same raw materials. In the Amazon basin, for example, the universal availability of toucan and parrot feathers and red vegetable dye gives an underlying unity to the local variation in ornaments and painted decoration. Another homogenizing factor is the adaptive superiority of certain cultural traits under given climatic conditions. Clothing is one example. In the humid heat of the Amazon basin, anything that interferes with the free circulation of air over the skin retards heat loss and consequently hinders normal physiological processes. The fact that of the five tribes in our sample only the Jívaro wear clothing, and even they frequently remove most of it when engaged in physical activity, is a cultural response to this biological fact.

Housing is another trait closely linked to the environment. Although the annual temperature fluctuation is minimal in the Amazon basin, the daily fluctuation is large enough to produce discomfort. Physiological characteristics adaptive to daytime heat, such as lower body temperature and higher insensible perspiration, are a disadvantage in the early morning hours when evaporation is reduced and temperature drops below 65°F. Amazonian peoples have generally met this problem by constructing large high-roofed dwellings with tightly thatched walls and roof, which provide shade during the heat of the day and insulation against the cold night air. Daily temperature fluctuation inside is also reduced by eliminating windows and providing only two small doors. Fires are kept burning all night near the sleepers to provide additional warmth. The heat radiated by a large number of human bodies probably also helps to maintain a comfortable nighttime temperature.

Another trait that has become almost universal because it satisfies a biological need is the manufacture of mildly alcoholic beverages from manioc, sweet potatoes, or seasonal fruits. The Jívaro are notable in consuming their major staple food—sweet manioc—as a slightly fermented drink rather than in solid form. Such beverages play a significant adaptive role in a hot humid climate, where continual perspiration is necessary to maintain normal body temperature. The resulting moisture loss is too great to be made up by drinking water, even with major conscious effort. The existence of more palatable beverages, however, not only encourages consumption of the required amount of liquid but also supplies vitamins and calories.

Beyond these general climatic features, the terra firme environment has several other characteristics that exert important effects on human exploitation. Predominant among them is the infertility of the soil, which limits the intensity of agricultural use. Another is the combination of high humidity and high temperature, which makes storage of most foods difficult except for short periods. A third is the low concentration of both plant and animal protein, so that a balanced diet can be obtained only through utilization of a considerable variety of subsistence resources.

It is obvious that no human community can survive unless its members obtain not only sufficient calories to sustain life but also a certain level of proteins, vitamins, and minerals. The fact that the Amazonian lowlands were well populated at the time of the European conquest indicates that a successful cultural adaptation to its special characteristics had been achieved. Such an adaptation would have to maintain a balance between population density and the long-range carrying capacity of the environment to prevent overexploitation and a consequent irreversible depletion of essential resources. This could be accomplished in two principal ways: 1) by developing measures to maximize the food return from a given area; and 2) by preventing an increase in population size or concentration sufficient to endanger the resources of the local environment. When the customs and beliefs of the five tribes in our sample are examined from this point of view, significant adaptive aspects become obvious.

Techniques for Maximizing Food Return

Although the same major food resources are potentially available throughout the Amazon basin, the five sample tribes differ in what they consider to be edible, the proportion of various foods that is consumed, and the way in which they are prepared. Since none of the populations exhibits signs of nutritional deficiency, it is fair to assume that these variations represent alternative solutions to the problem of maintaining a balanced diet and a steady food supply.

Although slash-and-burn agriculture is universal, the staple plants are not the same. The Sirionó, Jívaro, and Kayapó raise sweet manioc, while the Waiwai and Camayurá prefer bitter manioc. All consume large quantities of sweet potatoes, and the Kayapó depend on this plant more heavily than on manioc. The predominance of root crops in the list of cultigens has a simple environmental explanation. Manioc, sweet potatoes, and yams

Waiwai man burning a new field clearing. Although the slash has been drying for about six weeks, it is still too wet to burn well.

Waiwai communal house in a small clearing. The lean-to is used for most domestic tasks because the house interior is dimly lighted. Manioc and banana plants are growing in the garden behind the house.

Waiwai woman making a manioc grater by inserting small triangular stone chips into a board. The painted area will ultimately be filled.

A Waiwai family group inside the house. The husband takes care of the baby while resting in his hammock. The one beneath belongs to his wife. Arrows, ornaments, and other possessions hang from the thatch.

Young Waiwai men sitting around a babricot where meat is roasting. The small wooden stools are carved and are used exclusively by men.

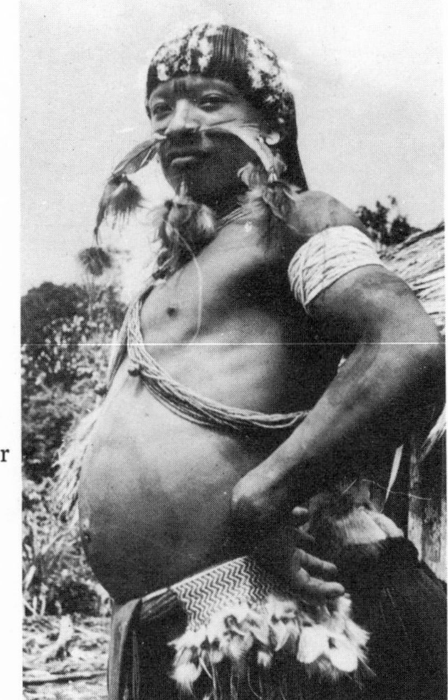

Waiwai man dressed for festival.

Typical Jívaro house, located on a hilltop for defense. The narrow door at the end is another defensive feature.

Jívaro woman making pottery by the coiling technique.

Jívaro man spinning cotton inside the house. The wrap-around skirt is typical male attire. A platform bed with a footrest is visible against the wall. Next to it (left) is a loom. The pottery jar contains the staple drink made from fermented sweet manioc.

Camayurá family resting in their hammocks while a wafer of manioc (cassava) bread bakes on a griddle over the fire.

Newly planted manioc field. Stumps and unburned trunks are interspersed with manioc cuttings, which look like slanting bundles of sticks.

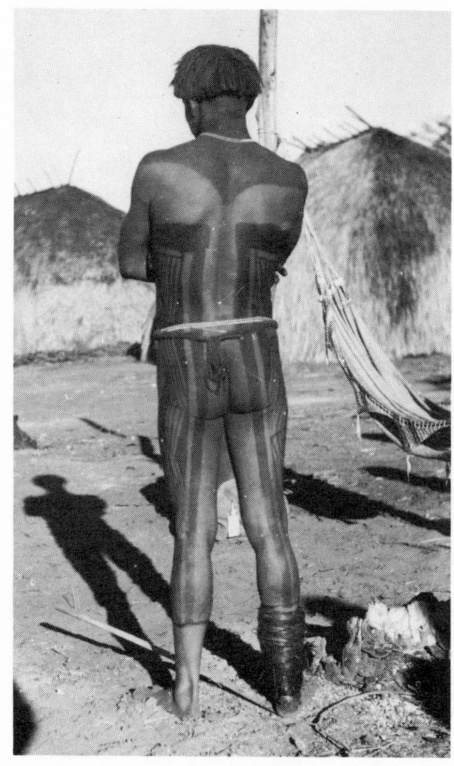

A Camayurá man with typical body decoration of black paint. He faces two of the communal houses that form the village circle.

The clear water of the Tapajós (foreground) mixes with the muddy water of the Amazon (left) opposite the town of Santarem.

Inundated forest on the varzea near the mouth of the Rio Negro during August when the water level has begun to fall.

The Rio Napo, a western tributary of the Amazon, near the end of the dry season when large sand bars appear in the riverbed.

Typical river in Guiana region of northeastern Amazonia late in the dry season. At high water, the beach in the foreground and the opposite bank are submerged.

are tolerant of poor soil, resistant to disease, and insensitive to extremes or irregularities of rainfall, with the result that they produce a much greater return for the labor invested than other kinds of plants. Furthermore, they bear continually and the tubers can be left in the ground until needed, eliminating the spoilage that accompanies the storage of a seasonal harvest. Although relatively low in nutrients, their food value can be enhanced by special methods of processing, particularly if fermentation is encouraged. Both the Jívaro and the Waiwai have utilized this potential. Only the Jívaro, however, consume the most nutritious portion of the plant, which is the protein-rich leaf. Another cultigen high in protein exploited only by the Jívaro and the Camayurá is the peanut.

Although maize is not well suited to the soil of the terra firme, it is grown to some extent by four of the five tribes in our sample, the only exception being the Waiwai. It is eaten both roasted on the cob and in the form of meal, but its seasonality and the adverse conditions for storage of ripe grain make it an important component in the diet during only about two months of each year.

Protein requirements are met principally by the consumption of animals, birds, reptiles, amphibians, fish, and turtle or reptile eggs. Except for the Sirionó, who look upon all fauna except insects as edible, there is a tendency to concentrate on a portion of the available resources rather than to utilize them all. To some extent, this is a reflection of unusual local abundances of a particular kind of food. The best example is the Camayurá, who ignore game in favor of fish, which can be harvested intensively in nearby lakes without danger of depletion.

While the other four tribes utilize fish to some extent, they are primarily consumers of meat. This creates a problem because the intensive hunting of animals and birds leads inevitably and rather rapidly to the exhaustion of the local supply. Other things being equal, however, the smaller the community, the more gradual the effect and the longer the food supply will last. Since sedentary life has superior survival value over wandering, the best adaptation is the one that permits the maximum degree of village permanence consistent with conservation of the natural resources. In other words, the village must be moved frequently enough to prevent irreversible damage to flora and fauna, but not so frequently that the potential advantages of sedentary life will be unnecessarily sacrificed. Since most terra firme tribes move their villages on the average every five years, this period probably represents the optimum permanency under typical ecological conditions.

The Kayapó have developed a different solution to the problem. Instead of moving the settlement periodically, they move the community for part of every year to a different hunting range. Thus, in the immediate vicinity of the village, the pressure on the game is relieved sufficiently to maintain an equilibrium, or at least to decrease significantly the speed at which depletion takes place. The large community size and cultural elaboration sustained by the Kayapó present a marked contrast to the level achieved by the Sirionó, who follow a similar pattern of alternation between rainy season sedentariness and dry season wandering. It seems likely that the inferior quality of the Sirionó environment is principally responsible for this contrast.

The list of edible wild seeds, nuts, fruits, berries, roots, and other vegetable products in the terra firme forest is very long, and no Amazonian tribe has ever approached complete exploitation of them all. Among the factors that influence the composition of the annual subsistence round of every group are: 1) the abundance of the seasonal harvest; 2) the absence of conflict between the time of ripening of wild foods and essential agricultural activities; and 3) the contribution of each kind of food to the overall nutritional balance. While the first two considerations may be evident to the people involved, the third could have exerted its effect only through natural selection over a long period of time.

An examination of the diagrams showing the annual subsistence round for each of the sample tribes (Figs. 8, 10, 13, 14, 16) reveals clearly the complementary relationship between the exploitation of wild nuts and fruits and the utilization of other sources of food. The Kayapó dependence on fresh Brazil nuts comes in August and September, for example, when garden produce is in low supply, while the harvest of mature fallen nuts takes place between December and February, before new crops are mature. Brazil nuts are not only obtainable in quantity and well suited to storage, but are extremely rich in protein and other nutrients. Another plant that is both highly nourishing and available during the time that gardens are unproductive is the piqui. The Camayurá have improved the yield of this fruit by planting trees, which they continue to exploit after the village has moved to a new location. Similar kinds of selections have been made from the potential faunal resources; seasonal feasting on turtle eggs by the Camayurá is an outstanding example.

Another variable aspect of the annual subsistence round is the number of foods that are important in the diet. The Waiwai, at one extreme, concentrate heavily on wild game and bitter manioc, which they prepare in a great variety of ways. Fish and several other root crops are also emphasized, but minimal use is made of wild plant foods. The Kayapó show a similar primary dependence on wild game and a root crop, in this case the sweet potato. During March and April, maize is consumed in quantity. In addition, however, they depend very heavily on wild plants, especially Brazil nuts, palmito, and various fruits, and on tortoises. The Jívaro exhibit still another pattern, which concentrates on domesticated and wild foods available the year round, with secondary emphasis on seasonal cultigens, such as maize and chonta. More complete information on the subsistence round of each group would certainly reveal additional variations that could be attributed to the specific conditions of the local habitat.

In spite of the climatic handicaps, techniques have been developed for preserving some kinds of food for short periods of time. The most reliable method for fauna is to keep the creature alive. This is impractical in most cases, but the turtle is an important exception. Turtles are typically stockpiled either for consumption during feasts or as insurance against the days when the hunters return empty-handed. In other cases when the catch is too large for immediate use or a surplus of meat is required for a forthcoming feast, it is preserved by roasting over a smoky fire and then wrapped in leaves.

Drying and fermenting are two techniques often employed for the preservation of certain kinds of plant foods. Bitter manioc roots can be stored for some time if they are well dried; alternatively, the grated pulp can be dried and stored. Boiled piqui pulp is wrapped in leaves by the Camayurá and submerged in water, where it keeps for months. Such practices do not significantly increase the reliability of the food supply, however, or prolong the utilization of plants much beyond their natural period of availability. Subsistence security in the terra firme environment stems instead from the year-round productivity of root crops, such as manioc and sweet potatoes, and the daily availability of wild game or fish.

Techniques for Control of Population Size

In order for an annual subsistence round to develop and be perpetuated, the resources that enter into it must remain available indefinitely at a more or less constant level of productivity. The existence of such rounds implies, therefore, that an equilibrium has been achieved between the population and the environment, so that intensity of utilization does not exceed the regenerative capacity or rate of replacement of the resources consumed. Since such an equilibrium would be threatened if man's intrinsic reproductive capacity were allowed full expression, natural and cultural measures must operate to inhibit multiplication beyond an optimum level and to distribute the existing population in such a way as to take maximum advantage of the carrying capacity of the environment. Many cultural practices that seem peculiar or even cruel to civilized observers can be explained in terms of their relevance to one or both of these functions.

At this point, it is worth digressing to make a distinction between the purpose of a trait and its function. The purpose is frequently evident to its practitioners because of the positive or negative sanctions involved. A Kayapó boy, for example, knows that he must submit to ordeals in order to increase his courage and endurance. A Jívaro man is aware that he must avenge the death of a relative in order to preserve the wellbeing of the remaining members of the family. A Camayurá kills twins at birth because they are a sign that the spirits are angry with the parents for having broken a taboo. A Waiwai mother abstains from meat eating so that her young infant will grow strong. The same kind of trait may have a different purpose in different tribes. The purpose of a ceremony, for example, may be to assure an abundant harvest, to honor or pacify the souls of the dead, or to strengthen the bond between the tribe and its guardian spirits; the purpose of warfare may be to obtain captives or to avenge a wrong; the purpose of cannibalism may be to insult the victim or to acquire certain admirable personal qualities, such as courage.

The function of such behavior, on the other hand, may be totally different, and is seldom evident to the persons involved. It may be obscure even to an anthropologist because of the complexity of interaction between traits or the incomplete nature of his ethnographic information. If it is a valid assumption, however, that any population must achieve an equilibrium with the carrying capacity of its environment in order to continue to exist indefinitely, then cultural behavior that has a limiting effect on population size and concentration can be assumed to have evolved in order to fulfill this function. Ironically, the more effective the culture is in overcoming natural sources of attrition, such as starvation, disease, accident, or malnutrition, the more important it becomes to develop cultural substitutes unless concomitant advances in the subsistence technology make a larger population density possible.

Even as far back as the early Pleistocene, human population size was either stabilized for long periods of time or increased at a rate well below that permitted by natural human fertility, implying the existence of cultural measures leading to attrition. This inference is based on two facts. First, although a woman may be expected to give birth to an average of eight children during her reproductive years, families of this size are rare except under disturbed cultural situations, such as now exist in underdeveloped areas where modern medicine has removed formerly stringent natural controls. Second, four births per female are sufficient for the maintenance of a steady population size, assuming that two individuals will die before maturity and leave the other two to replace their parents and in turn have children.

The primary adaptive problem facing a culture that must maintain a stable population size is consequently how best to inhibit the natural rate of increase without sacrificing the resiliency provided by the reproductive capacity. The potential number of effective solutions is very large, not only because many different kinds of measures can be employed, but also because attrition can be intensified at different stages of the life cycle. Thus, if the productivity of the subsistence resources is so low that each person must be largely responsible for supplying what he consumes, few nonproducers can be tolerated and measures that concentrate on preventing conception or eliminating a certain proportion of the newborn are most economical. If children perform a useful function or do not constitute a serious burden, attrition measures can be delayed until puberty. Other controls operate on adults, either by reducing longevity or interfering with normal procreation. Each of the tribes in our sample employs several of the potentially available methods of population control, but the number of measures and the intensity of their application differ considerably from one group to another.

The inverse relationship in intensity that exists between natural and cultural means of population size control becomes evident when the Sirionó are compared with the other four terra firme tribes (Table 2). Although the Sirionó alone lack effective cultural controls at the prenatal and juvenile levels, the average number of children per family is only two. The obvious implication is that natural controls reducing the birth rate are so strong that cultural measures are superfluous and, in fact, there are reports that miscarriage is relatively common.

At the adult level, on the other hand, the Sirionó have one custom not represented among the other groups, namely the abandonment of individuals too ill or infirm to move with the band. To understand this behavior, it must be remembered that starvation threatens unless all adult members of the community are able-bodied. Injury or illness severe enough to prevent participation in the food quest places the individual in jeopardy because other members of the band cannot provide him with the nourishment or rest necessary to his recovery without endangering their own survival. In such circumstances, infirmities that might otherwise not cause death are likely to be fatal. Recovery is also impeded by the patient's feeling of hopelessness, which is unalleviated either by medical treatment (the Sirionó have no shamans), or by sympathetic attention from relatives. While their indifference serves to shield the survivors from psychological damage resulting from exposure to frequent deaths, it intensifies the patient's anxiety and consequently promotes his demise. Furthermore, since loss of appetite is interpreted as a sure sign of the seriousness of an illness, a patient will force himself to eat in spite of pain or gastro-intestinal symptoms, often with deleterious consequences. It is inconceivable that behavior so detrimental to the individual would be perpetuated unless it were vital to the long-term survival of the group. The expendability of older members of the community is reflected in the absence of any role in Sirionó culture that rewards the increased experience or knowledge that comes with age. Quite the contrary, status is achieved by skill in the food quest and thus depends on strength, agility, and other qualities of youth.

The Waiwai, Jívaro, Kayapó, and Camayurá all observe a vari-

Table 2. Cultural mechanisms of population control among terra firme groups

POPULATION SIZE CONTROLS	Sirionó	Waiwai	Jívaro	Kayapó	Camayurá
Prenatal					
Sex taboo	♀ 1 mo. after childbirth	♂ 1 mo. while preparing curare	♂ 3-6 mo. after taking head ♂ & ♀ while child nursing (2-3 yrs.)	♀ prenatal until child walks	♂ few days before wrestling or dancing
Contraception			Yes		
Abortion			Rare	By unmarried ♀ and wives before 1st child	If conceived while nursing (3-4 yrs.)
Juvenile					
Infanticide		If unwanted by ♀ 5th child of same sex	If deformed Father from another tribe	If twins If ♀ dies	If deformed If twins Born while ♀ nursing ♀ unmarried
Adult					
Blood revenge		By magic	By murder		
Death penalty		♂ in adultery	♀ in adultery		
Warfare			All killed but young ♀s	Yes	
Abandonment of ill or infirm	Mandatory	Permissible			
POPULATION DENSITY CONTROLS					
Warfare			Yes	Yes	Yes
Sorcery		Yes	Yes	?	Yes
Fear of "outsiders"			Yes	Yes	Yes
Love of freedom		Yes	Yes		

ety of other customs that affect population growth either by preventing births or by eliminating a certain proportion of the newborn. Contraception and abortion are employed, especially by the Kayapó, but the most common method is to place restrictions of varying degrees of intensity on sexual relations. While the periods of abstinence observed by the Waiwai and Camayurá seem too ephemeral to be effective as population control measures, those of the Jívaro and Kayapó are of considerable duration. A Kayapó woman must remain sexually continent from the onset of pregnancy until the child is able to walk, assuring that a sibling will not be born before the previous child is ready to be weaned. Since the Kayapó practice monogamy, observance of this taboo by the wife also inhibits the procreative capacity of the husband. On the other hand, where polygyny is characteristic and female adultery is punishable by death, as among the Jívaro, a taboo observed by the husband simultaneously removes two or more females of child-bearing age from the reproductive role. The requirement that a warrior observe continence for three to six months following the taking of a head becomes intelligible in this light. In addition, a Jívaro woman is obliged to refrain from sexual relations while her child is nursing, which it does for two or three years.

Prenatal controls are supplemented by selective infanticide. Deformed children are killed (except by the Sirionó) because of the belief that they have been fathered by evil spirits. While this practice can be understood as a way of relieving a society of the burden of feeding individuals potentially incapable of becoming useful members, it has another less obvious adaptive significance. It has been shown that the rate of deformity increases when the mother is supplied with an inadequate diet during pregnancy. Defective maturation of the fetus is consequently likely to be a sign that the existing population is at a dangerous level with regard to the food supply. Under such circumstances, the elimination of deformed infants not only increases the mother's chance of survival by removing the strain of nursing but also helps to rectify the imbalance that has developed between population size and subsistence resources.

The absence of an alternative food for human milk makes infanticide a humane substitute for lingering death by starvation when the mother dies. The inability of a woman to nourish more than one infant also underlies the Camayurá practice of killing any child born while the previous one is still nursing. This practice also assures a spacing of three to four years between offspring.

Among the Kayapó and Camayurá, sexual taboos, abortion, and infanticide are apparently sufficient to keep population increase within acceptable limits. Among the Waiwai and to an even greater extent among the Jívaro, however, several other important cultural practices exist that have the effect of eliminating adults, and particularly males, from the breeding population. These include blood revenge, warfare, and the punishment of adultery by murder.

The Jívaro have elaborated blood revenge and warfare to a point where these activities set the tone for the whole society. Boys from early childhood are encouraged to develop the traits and attitudes that will make them successful warriors, and their status as adults depends on their skill and accomplishments. The lure of prestige is reinforced by the knowledge that supernatural sanctions will follow any failure to avenge the death of a relative. The consequence is a high male death rate and an adult sex ratio of one male to two females.

One of the most interesting aspects of Jívaro culture is the manner in which this imbalanced sex ratio has been made an integral part of the adaptive complex. Since women assume the primary subsistence role, the loss of half the adult male population does not threaten the food supply as it would if the sex roles were more equal. Furthermore, in contrast to the Sirionó hunter who spends 50 percent of his time in search of game and often returns empty-handed, a Jívaro man is able to keep his family supplied with meat with an expenditure of less than 20 percent of his time. The discrepancy in productivity is further enhanced when it is recalled that a Sirionó man feeds one wife and a couple of children, whereas a Jívaro not only supports two or more wives and their offspring but also provides a surplus of meat for consumption by guests on numerous festive occasions. The situation is the opposite for women, since the preparation of the principal Jívaro dish is a time-consuming task involving the mastication of sweet manioc and there is a limit to the amount of time that a woman can devote to this job and still fulfill her other domestic obligations. Under these circumstances, the most practical way to assure adequate production is to increase the number of females, who are producers as well as consumers, relative to males, who are only consumers and who compete in this role with children. The Jívaro subsistence pattern thus not only makes possible the elimination of a certain number of adult males, but actually operates more efficiently with an unbalanced sex ratio.

The existence among the Kayapó of a warrior class, composed of young men at the height of their physical activity and consequently of food consumption but who make little or no contribution to food acquisition, represents an alternative means of dealing with surplus adult males in the population. A member of this class obtains his food from his mother or, if married, from his wife. His primary responsibility is to keep fit and to engage in warfare. Since the subsistence burden rests on women and male heads of families, these young men can go off on military expeditions without affecting the village food supply. Although the Kayapó have a reputation for warfare equivalent to that of the Jívaro, the low male mortality reflected in an approximately equal adult sex ratio implies that its function is not primarily control of population size.

A significant aspect of most postnatal measures of population control is the random manner in which they are applied. Such randomization maintains the sex and age ratios appropriate to the functioning of the culture, and prevents the elimination of too large a proportion of the potentially most useful members of the society. When the circumstances under which infanticide is practiced are considered from this point of view, it becomes evident that except in the case of deformed children, the criteria are irrelevant to the health, intelligence, birth order, or sex of the victim. Similarly, the use of divination by a Jívaro shaman to identify the person responsible for causing a death allows the choice to fall on an adult without regard to age, status, or ability, or even to friendship, kinship, or hostility for the deceased. Although a male is usually designated as the culprit, revenge can be also satisfied through taking the life of one of his female relatives.

Warfare among the Kayapó and Jívaro differs strikingly from that of our own society in the manner in which warriors are chosen. Whereas we select the strongest and most intelligent males for potential slaughter, these more primitive peoples make no such discrimination. Among the Kayapó, all boys become members of the warrior class at a certain age. All Jívaro men are under heavy social pressures to execute blood revenge or to take a head, and thereby to risk their own lives. Other things being equal, it is probable that such circumstances favor survival of the stronger and more able rather than otherwise. The same is true of the victims. The goal of Jívaro warfare is extermination of the entire population of the enemy village, so that superior intelligence, agility, and strength will enhance the chances of escape.

Techniques for Control of Population Density

While a great variety of wild plant and animal foods is potentially available from the terra firme environment, the distribution of these resources over the landscape is typically thin and scattered. Furthermore, the low fertility of the soil makes frequent moving of gardens necessary to maintain a reasonably high level of productivity. Under such circumstances, a larger population can exist indefinitely without irreversibly depleting the local food resources if it is dispersed than if it is concentrated. It thus comes as no surprise to discover that cultural mechanisms exist among the terra firme tribes to prevent population concentration. On the other hand, human beings, like other social animals, require a minimal population density for the satisfaction of certain kinds of social and psychological needs. A balance must be achieved, therefore, between optimum density in terms of the ecological situation, which is relatively low, and optimum density for the satisfaction of basic social needs, which may be somewhat greater.

In all of the sample societies, the minimum economic unit is the extended family. Since the division of labor is along sex lines, members of this group possess jointly all of the knowledge and skills required to provide the food, shelter, and equipment necessary for survival and reproduction. The extended family consequently represents the smallest feasible independent segments, and in two of the societies in our sample, the community organization is based on this minimal unit. Each Waiwai and Jívaro village consists of a single communal house separated from its nearest neighbor by several miles of uninhabited forest. The existence of strong sanctions not only against the clustering of households but also against enlargement of a single household beyond a certain size implies that a higher population concentration is incompatible with the long-term exploitation of the habitat.

The major factor in the perpetuation of a dispersed community pattern among the Jívaro and Waiwai is a pronounced fear of sorcery. In both societies, personal security is assured only within the extended family. Anyone else is potentially dangerous, even if he is a relative. Although sorcery can be practiced from a distance, proximity increases the hazard, and the larger the household group, the greater the risk of a conflict serious enough to arouse deadly enmity. Consequently, households tend to subdivide when the number of occupants exceeds psychologically comfortable limits. Among the Waiwai, a split may occur when the village moves to a new location. Among the Jívaro, one of the constituent nuclear families withdraws to found a new household after several children have been born. The significant aspect of this situation is the severity of the penalty for population concentration. In fact, a potential death sentence hangs over every member of a household that increases beyond optimum size.

Some of the same sanctions that inhibit the enlargement of any individual settlement also prevent settlements from clustering. Sorcery is again a major repelling force, not only because its danger tends to increase with the accessibility of the sorcerer to his victim, but because a larger population increases opportunities for conflicts to arise leading to this kind of revenge. Both the Waiwai and the Jívaro resist being drawn into quarrels that develop between other members of the household. Involvement is dangerous because if the matter is not resolved satisfactorily, one or both parties may resort to sorcery. It is relevant in this connection that both cultures place a high value on personal liberty. Karsten (1935, p. 269) says of the Jívaro that "their unbounded sense of liberty and their desire to be independent . . . is one of the reasons why they do not live in villages but each family separately, for in this way conflicts are more easily avoided." Fock (1963, p. 233) uses similar language with reference to the Waiwai: "The individual Waiwai has the greatest objection to getting mixed up in the conflicts of his village kinsmen, even when his sympathies are clearly on the side of the one party. His attitude is dictated by fear of provoking fatal blowing [sorcery], and a strong wish to avoid outside interference in his personal freedom, should he find himself in a similar situation."

Although the Camayurá do not practice sorcery, they fear that it may be directed against them by members of neighboring tribes. As a consequence, intertribal gatherings take place in an atmosphere of tension that seems at first glance to be incompatible with the amount of economic and social involvement that exists. In contrast to the Kayapó, Sirionó, Jívaro, and Waiwai, who have little or no significant dealings even with members of related communities to say nothing of alien tribes, the Camayurá not only depend on non-Camayurá for several kinds of essential articles, such as pottery, but also invite them to participate in their major festivals. The advantages of this intertribal cooperation must be balanced against the dangers of overexploitation of local resources, and the fear of sorcery is an effective method of preventing the concentration of too many people in one place for too long a time (the optimum duration being determined by the amount of strain placed on local food resources). In Camayurá society, the kinship bonds resulting from intermarriage with members of other tribes inhibit the intensification of hostility into a pattern of blood revenge that would jeopardize the beneficial aspects of intertribal relations. A large number of checks and balances of this kind permeate every culture that has achieved a successful adaptation to its environment.

Another regulator of population density among Amazonian tribes is warfare, and this function is evident in several features that distinguish it from warfare among more civilized groups. In the first place, raiding does not lead to the acquisition of land. On the contrary, there may be strong sanctions against territorial annexation. The Jívaro, for example, "fear and detest the country of their enemies, where secret supernatural dangers may threaten them even after they have conquered their natural enemies. . . . The land of the enemy is therefore abandoned as soon as possible" (Karsten, 1935, p. 278). Second, the most intense hostility is often directed at other communities belonging to the same tribe, rather than at members of other tribes. Karsten points out that "It is characteristic of the Jibaros that they especially wage war against tribes belonging to their own race and speaking the same language," and that the goal is "to completely annihilate the enemy." Similarly, Dreyfus (1963, p. 93) says that "not only does there exist no sentiment of tribal solidarity among the Kayapó, but even among related groups separation always leads to hostility." Finally, food and material goods belonging to the enemy are typically destroyed, and looting is never the motive for aggression.

In spite of the absence of material rewards, warfare is an integral part of both Kayapó and Jívaro culture. Both tribes inculcate a love of fighting into young boys by recounting tales of past injuries and exploits, by stimulating them to emulate brave warriors, and by exposing them to ordeals to develop their courage and endurance. In both societies, killing an enemy is prerequisite to the attainment of status and prestige by men. Among the Jívaro, the psychological pressures are even stronger, since a failure to execute blood revenge not only negatively affects the status of the man who thus shirks his duty, but exposes his whole family to retaliation by the spirits, resulting in crop failures, illness, or even death. Indeed, Karsten (1935, p. 259) has observed that "wars are to such a degree one with their whole life and essence that only powerful pressure from outside or a radical change in their whole character and moral views could make them abstain from them." Paradoxically, abandonment of warfare under acculturative pressures has decreased rather than enhanced the chances of survival of the tribe by undermining the intricate adaptive structure of the culture.

Another widespread custom that becomes intelligible in the context of the ecological imperatives of the terra firme environment is abandonment of the house on the death of an adult resident. Three tribes in our sample follow this practice with differing degrees of intensity. The Sirionó, who abandon the house at the death of any adult, are at one extreme, while the Jívaro, who do so only on the death of the household head, are at the other. Intermediate are the Waiwai, who take this action (at the present time) after the death of the chief, the shaman, or a wife of one of these men. Even when restricted to a few selected individuals, this custom frustrates any tendency to maintain a higher degree of sedentism than is compatible with the preservation of the ecological equilibrium. Among the Sirionó, the greater intensity of application might at first glance seem redundant in view of the already high mobility of the band during a major part of the year. The total absence of sanitation during the sedentary period, however, probably makes abandonment of the house a practical health protection measure.

SUBSISTENCE EMPHASIS AND SEXUAL DIVISION OF LABOR

While the division of labor by sex has biological roots, it has been elaborated during cultural evolution as a device for distributing the work load as equitably and efficiently as possible. Where biological differences make one sex clearly superior to the other in some activity, as in child care and tasks requiring physical strength, the same allocation is practically universal in human societies. As culture has grown in complexity, however, sexual division of labor has become one of the most variable areas of social behavior. An examination of the sexual division of labor among the five sample terra firme tribes shows that allocation of activities seemingly unrelated to subsistence is an integral part of the adaptive configuration.

When the tasks allotted to men and women are tabulated and compared (Table 3), two factors are immediately evident. In the first place, essentially the same kinds of activities are performed by the Camayurá, Kayapó, Jívaro, Sirionó, and Waiwai. Although major differences exist in intensity, seasonality, and the relative importance of the various tasks, subsistence is based on similar combinations of wild and domesticated foods, arts and crafts involve the same categories of raw materials and finished products, and daily household activities follow a similar routine. The largest disparities occur in the presence or absence of ceremonial activity and warfare.

The second obvious feature is a lack of consistency in the tasks assigned to men and women. Except that fishing and hunting are always done by males and domestic tasks are performed principally by females, the allocation of tasks appears at first glance to be arbitrary. In arts and crafts, the Jívaro and Sirionó are at opposite poles, since most of the tasks performed by Sirionó women are done by Jívaro men. The pattern among the Kayapó and Waiwai is similar to that of the Jívaro, but the Camayurá tend to be intermediate, allocating to women several duties performed elsewhere by men. Since a culture is an integrated system, these differing patterns must reflect differing emphases in other activities. As a matter of fact, they are clearly related to the roles of the sexes in subsistence.

The effect of the subsistence pattern on the sexual division of labor in arts and crafts is most obvious among the Waiwai and the Jívaro. In both groups, women spend approximately 50 percent of their time in processing manioc. Jívaro women are also responsible for planting and harvesting the staple crops. When these time-consuming daily activities are added to caring for children, fetching water, and other recurrent domestic tasks, they fill

Table 3. Division of labor by sex among terra firme groups. (X) = minor contribution

Tasks	Sirionó ♂	Sirionó ♀	Waiwai ♂	Waiwai ♀	Jívaro ♂	Jívaro ♀	Kayapó ♂	Kayapó ♀	Camayurá ♂	Camayurá ♀
SUBSISTENCE										
Hunting	X		X		X		X			
Fishing	X		X		X		(X)		X	
Gathering	X	X	(X)		X	X	X	X	X	X
Agriculture										
Clearing and burning	X	X	X		X		X		X	
Planting staples	X	X	X	(X)		X		X	X	
Weeding			X	X		X	X	X	X	
Harvesting staples	X	X	X	X		X		X	X	X
DOMESTIC										
Getting water		X		X		X		X		X
Collecting firewood		X		X	X			X	X	X
Child care		X	X	X		X	(X)	X		X
Processing manioc				X		X				X
ARTS AND CRAFTS										
Spinning cotton		X		X	X			X		X
Weaving cotton			X		X					X
Making twine		X	X		X		X			X
Weaving hammocks		X	X							X
Baskets and mats		X	X		X		X		X	
Wood carving	X		X		X		X		X	
Gourd utensils	X		?		?		X	(X)	X	
Musical instruments			X		X		X		X	
Feather ornaments		X	X		X		X		X	
Seed, cotton, etc., ornaments		X		X	?	?		X	X	X
Pottery		X		X		X				
Dugout canoes			X		X				X	
CEREMONIES			X	(X)	X	X	X		X	X
WARFARE					X		X			

a woman's day. A man's subsistence duties, on the other hand, require less than 20 percent of his time, leaving 80 percent for other kinds of activities. Under such circumstances, it is not surprising to find that men manufacture most of the articles used by both sexes.

A similar correlation underlies the Sirionó division of labor, although here the sexual roles are reversed. Hunting, which is a primary subsistence activity, consumes at least 50 percent of a man's time, and men share equal responsibility with women in other major subsistence tasks. Sirionó women, on the other hand, do not have to spend significant amounts of time in food preparation, and consequently are able to take on the manufacture of baskets, hammocks, and other articles without undue hardship.

The division of labor among the Camayurá is also correlated with the differential roles of the sexes in subsistence. Here, the men are the main providers of both wild and cultivated foods, while women are charged with their preparation for consumption. Once again, the processing of bitter manioc is a continuing and time-consuming task. The more even distribution between the sexes of arts and crafts reflects the fact that the subsistence burden is shared more equally than is the case for other groups in the sample.

The correlation between subsistence contribution and the division of labor in arts and crafts is less clear-cut among the Kayapó, probably because their greater complexity of general cultural development tends to obscure the relationships between different kinds of activities. Since men manufacture most of the material goods, it should follow that their contribution to subsistence is less time-consuming than that of women. The data do not suggest an obvious imbalance, however. Although women perform most of the agricultural work aside from clearing the fields, this is a seasonal activity; nor is food preparation as laborious as for Waiwai, Jívaro, or Camayurá women, since bitter manioc is not a staple crop. Close examination of Kayapó arts and crafts reveals another relevant factor, however: several complex activities, including cloth and hammock weaving, and pottery and canoe making, are not represented. Many of the articles made by the men are relatively durable and consequently do not require constant replacement. As a matter of fact, Kayapó men probably devote more time to warfare and ceremonies than to carving wood and weaving baskets. Thus, while the quantitative data that would establish the exact ratio of male and female input in various kinds of activities are not available, the existing evidence is sufficient to indicate that the pattern of sexual division of labor is here again a reflection of the allocation of subsistence responsibilities.

INCIPIENT DIFFERENCES IN CULTURAL COMPLEXITY

A tabulation of the features generally considered indicative of level of cultural development shows that the Kayapó and Camayurá possess in incipient form several advanced features that are absent from the other three tribes. Both have a permanent multihousehold chief and a village council that meets nightly to discuss matters of community interest and concern. The Kayapó, who have settlements exceeding 500 inhabitants, also possess several other kinds of social groupings that cross-cut kinship lines and thus provide social integration between individuals unrelated by blood or marriage. Since one of the principal differences between the social organization of civilized and precivilized societies is the substitution of civil ties for kinship ties, this development among the Kayapó is of special interest.

The Camayurá exhibit in incipient form two other characteristics often mentioned in connection with the emergence of civilization: namely, occupational division of labor and formalized exchange via a public market. These practices promote the circulation of objects both within the village and between villages belonging to different tribes. On the other hand, the Kayapó, whose villages are much larger than those of the Camayurá, have maintained the characteristic Amazonian pattern of division of labor along sex lines and minimal informal exchange of nonsubsistence products between members of the community. Such discrepancies underline the fact that cultural evolution does not proceed uniformly in all aspects of a cultural complex or at a constant rate.

When the five tribes in our sample are compared in terms of the presence or absence of advanced social and economic features, it is evident that they can be arranged in an order of increasing complexity (Table 4). The Sirionó are by far the most primitive, lacking all the traits listed except for the household chief (who because of the Sirionó residence pattern also constitutes a multihousehold chief). The Waiwai and Jívaro are not much more advanced; they possess shamans as part-time occupational specialists, engage in a minimum amount of intervillage or intertribal trade, and show incipient expression of multihousehold organization and occupational division of labor in arts and crafts. In contrast, the Kayapó possess seven and the Camayurá eleven of the significant features. The incipient development among the Camayurá of markets, occupational specialization in arts and crafts, and tangible representations of supernatural beings is particularly noteworthy.

There is no doubt that the lower level of cultural complexity exhibited by the Waiwai and Jívaro is correlated with the larger

Table 4. Cultural traits of evolutionary significance among terra firme groups

Traits	Siriono	Waiwai	Jivaro	Kayapo	Camayura
SEDENTISM					
Village population	± 80	± 25	± 40	150+	± 110
Village permanency	under 6 mo.	± 5 yrs.	± 6 yrs.	Indef.	± 10 yrs.
SOCIAL ORGANIZATION					
Household chief	X	X	X	X	X
Multihousehold chief					
Temporary (war)			X		X
Permanent				X	X
Village council				X	X
Non-kinship based associations				X	
PART-TIME OCCUPATIONAL SPECIALISTS					
Shaman		X	X	X	X
Arts and crafts					X
TRADE					
Within village					X
Between villages or intertribal		X	X		X
Formal market					X
SPECIALIZED STRUCTURES					
Men's house				X	
Flute house					X
Chief's house				X	
RELIGION					
Ceremonial posts (idols?)					X

number and greater severity of cultural mechanisms for controlling population size and concentration in these two tribes. In other words, the same ecological factors that make these constraints necessary also close the door to evolutionary advancement.

CONCLUSION

An analysis of the aboriginal cultural adaptation to the Amazonian terra firme environment brings out two basic facts: 1) population size and density are held within specific limits by a number of strongly reinforced cultural practices; and 2) within this limitation, interplay between the special features of each local environment and the configuration of preexisting culture has produced many variations on a single basic theme. Nowhere is variability more pronounced than in subsistence pattern. Although the same wild food resources and cultigens are available (with minor exceptions) throughout the area, no two groups combine the same ingredients in the same proportions. Whatever the annual subsistence round that has been adopted, however, all of the essential nutrients are supplied in the required amounts.

While subsistence differences do not appear to imply differential success in adaptation to the potential food resources, the composition of the annual cycle has a significant effect on the rest of the cultural configuration. Heavy dependence on seasonal wild food necessitates periodic abandonment of the village, usually with temporary fractionization of the community, and this has certain kinds of social consequences. The way in which subsistence tasks are allocated between the sexes influences the division of labor in other activities, as well as kinship obligations and the forms of marriage. In most of these relationships, the cause and effect is not linear so that it cannot be determined in the case of the Jívaro, for example, whether polygyny permitted a particular subsistence emphasis to evolve or whether the subsistence pattern favored the institutionalization of polygyny. Entering into and complicating the situation is the elaboration of warfare and its repercussions on both subsistence responsibilities and the ratio of adult males to females. Clearly, cultural adaptation is a complex process in which effects become causes in an endless circle of interaction that maintains a functional integration in the course of constant though usually imperceptible change.

With the exception of the Sirionó, the terra firme tribes appear to have enjoyed abundant subsistence resources and a generally easy life. In fact, they illustrate very well the idyllic existence that has led temperate zone observers to regard Amazonia as a paradise not fully exploited by the indigenous inhabitants. It is dangerous, however, to jump to the conclusion that plentiful game and productive gardens imply an unused potential. Rather, an understanding of the limitations inherent in the terra firme environment leads to the conclusion that the reverse is true, and that abundance is a reflection of the equilibrium adaptation achieved by the aboriginal inhabitants. It is significant in this respect that zoologists have also come to the conclusion that it is an illusion to think that animals are commonly far less numerous than their environments would permit. As we will see, the disastrous consequences of uncontrolled post-European exploitation provide an even clearer demonstration that Amazonia is a counterfeit paradise rather than a land of unrealized promise.

Chapter 4

ABORIGINAL ADAPTATION TO THE VÁRZEA

The várzea, like the terra firme, is a variable environment. Whereas the terra firme variation comes from rainfall, soil composition, and topography, however, the significant features of the várzea are differential susceptibility to inundation and unequal extent. Above the mouth of the Rio Negro, the width of the várzea averages only about half that along the lower Amazon. This means that, generally speaking, twice as much várzea is accessible from a given amount of frontage along the lower portion of the river. In addition, the high várzea above the Rio Negro tends to be inundated more frequently than the land farther east.

Unfortunately, none of the aboriginal cultures of the várzea has survived to be studied by anthropologists. In contrast to the terra firme, whose vastness made it relatively immune to interference by early European explorers, the várzea was compact, accessible, and vulnerable. As a result, the aboriginal cultural pattern had been completely destroyed within 150 years of its discovery, leaving only fragmentary and sometimes biased eyewitness accounts to provide details of its former character. Although six rather extensive descriptions exist for the period between 1542 and 1692, the data they contain are often vague or inconsistent. There is much variation, for example, in the names assigned to tribes or "provinces," making it difficult to decide whether two reporters are speaking of the same group. Uncertainties can sometimes be resolved by comparing distances or reference points, but the rarity of distinctive landmarks and the inconsistency in units of measurement (for example, days of travel as against leagues) often make a decision impossible. Furthermore, the accuracy of specific information must be evaluated in the light of our knowledge that few observers kept notes during their travels, that impressions were often engraved on their memory under the stress of battle, and that the temptation to embroider or exaggerate for personal glorification may have been strong.

In spite of their deficiencies, however, the early chronicles make it clear that population density and level of cultural development were considerably greater on the várzea than on the terra firme at the time of European contact. The Omagua on the upper middle Amazon and the Tapajós at the mouth of the river that bears their name are mentioned with sufficient frequency in different accounts that a general cultural description can be pieced together. More recent sources add details on subsistence practices, while archeology provides evidence on settlement pattern and material culture. Combining these sources of information permits reconstruction of the general level of cultural development achieved on the várzea in pre-European times, preliminary to examination of the interaction between culture and environment.

THE OMAGUA

Location and Environment. Between the Negro and the Japurá, the várzea has an average width of about 15 miles, only half of that prevailing along the lower Amazon. Because of this constriction, annual fluctuations in water level are great and a large portion of the flood plain is inundated every year. Annual rainfall is between 80 and 100 inches. Although precipitation occurs throughout the year, less than 6 inches per month falls between June and September.

In 1542, the várzea from the Japurá to about halfway between the Coarí and the Purus was occupied by the Omagua (Fig. 17). By 1690, they had migrated westward and were living between the Napo and the Putumayo. The first mission was established in the latter region in 1686 and 27 additional missions were founded among the Omagua during the subsequent decade. In typical Spanish fashion, these settlements also incorporated Indians from

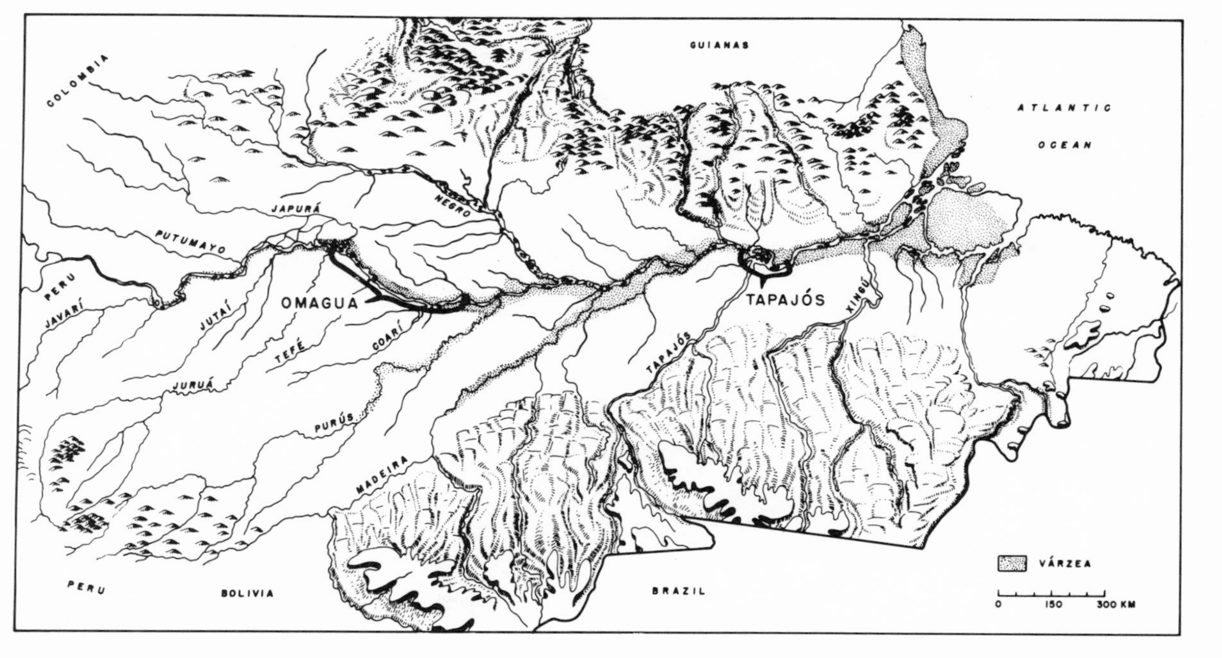

Fig. 17. Approximate extent of the várzea or Amazonian floodplain (stippled area). Carvajal, when he descended the Amazon in 1542, recorded the Omagua as inhabiting the region between the Japurá and midway between the Coarí and the Purus. He noted that the vicinity of the mouth of the Tapajós was thickly populated, but did not mention the inhabitants by name. (Geographical detail after Guerra, 1959, Fig. 7).

various other tribes, all of whom were indoctrinated in the Christian sacrament and the ways of civilized life. Their populations were constantly augmented by fugitives from downriver, who sought sanctuary from the brutal Portuguese slave raids. In 1710, however, particularly devastating raids penetrated the mission area, provoking its abandonment and the withdrawal of the survivors. San Joachim de Omaguas was reestablished below the mouth of the Ucayali, but had a population of only 522 in 1731. The Omagua language, which belonged to the Tupí linguistic family, was selected by the missionaries as the official medium for catechism and intertribal communication.

As a consequence of disease, slave raiding, and missionization, the Omagua way of life was nearly extinct by the beginning of the eighteenth century. We are fortunate, however, in possessing several documents that provide data on the aboriginal culture. Carvajal, who recorded the first descent of the Amazon in 1542; Simon, who passed through in 1560; Acuña, who descended the river in 1639, and Cruz, who followed in 1651, each contribute a few observations. Most of the details, however, come from the diary of Samuel Fritz, who served as the principal missionary among the Omagua from 1686 to 1723 and presided over their deculturation and decimation.

SETTLEMENT PATTERN. Settlements along the high banks were so close together, according to Carvajal, that "there was not from village to village [in most cases] a crossbow shot, and the one which was farthest [removed from the next] was not half a league away, and there was one settlement that stretched for five leagues without there intervening any space from house to house" (1934, p. 198). One village resembled a garrison and was located on a high spot overlooking the river. Another extended for more than five miles along the summit of a high bank, which was separated from the terra firme by a marsh. Large villages were divided into sections, each with its own landing place on the river. Roads frequently led toward the interior. A few settlements were located on the várzea, where they were accessible only by canoe during high water.

Omagua houses were large rectangular structures with cedar plank walls and palm thatch roofs. One village with a population of 330 consisted of a row of 28 houses, each occupied by an extended family. They were closely spaced and oriented with the long axis perpendicular to the river bank. There was a door at each end. The interior was kept cleanly swept, and was furnished with hammocks, large palmleaf mats, and many pottery vessels.

DRESS AND ORNAMENT. Cotton garments were worn by both sexes, a sleeveless shirt reaching to the knee by men and a very short wraparound skirt by women. Men usually left their shirts off, however, because they were less encumbered without them. The cloth was painted with multicolored designs.

The Omaguas were readily distinguished from other Amazonian tribes by their flattened foreheads. Fritz observed that shaping was done in infancy by "applying to the [babies'] forehead a small board or wattle of reeds tied with a little cotton so as not to hurt them, and fastening them by the shoulders to a little canoe, which serves them for a cradle" (1922, pp. 47–48). Acuña distainfully described the effect as "more like a poorly shaped bishop's miter than the head of a human being" (1942, p. 69).

SUBSISTENCE. An impression of subsistence surplus emerges from all of the early accounts. In one village, the members of Orellana's expedition found "a great deal of meat and fish and biscuit, and all this in such great abundance that there was enough to feed an expeditionary force of one thousand men for one year" (Carvajal, 1934, p. 192). Whenever they went ashore for provisions, they found large quantities of food. Acuña adds that "what is more amazing is the slight amount of work that all these things require, as we could observe daily from our own experience" (1942, p. 42). The staple was bitter manioc; maize also was important. Other crops were sweet manioc, sweet potatoes, peanuts, kidney beans, peppers, pineapples, avocados, and other fruits. Tobacco, achiote, gourds, and cotton were also grown. Although the chonta palm is not mentioned, the dense stands observed early in the nineteenth century along the middle Amazon suggest intensive cultivation in aboriginal times.

Fritz provides a few details on agricultural practices:

> The plantations . . . which furnish their sustenance, and the houses or ranches are generally situated on islands, beaches or banks of the River; all low lying lands liable to be flooded. . . . In order that there should be no lack of food during the season of the high-flood, which begins in March and lasts till June, they make a practice of harvesting the fruit of their new plantations in January and February (1922, p. 50).

This method produced as much manioc in four to six months as could be raised on the terra firme in a year and a half. Acuña observed that the annual inundation of the fields "fertilizes them with its mud, so that the soil never becomes sterile even if year after year, it is required to produce maize and manioc, which is their staple food and which exists in great abundance" (1942, p. 35).

Maize was stored in the houses or in elevated granaries. Sweet and bitter manioc tubers were harvested before inundation and interred in pits on the várzea, where they remained until the water level dropped. Carvajal reports that they could be kept for two or more years, and "although this yuca and mandive may rot, when pressed it becomes better and of greater sustenance than when fresh, and from it they make their drinks, flour, and cassava bread" (1934, p. 50).

Bitter manioc was made into "biscuit" or cassava bread. Carvajal explains that although "this biscuit will seem odd to those who do not know about it or have never seen what it is, not being made of wheat flour, it must be pointed out that there the Indians had great quantities of large cakes made out of cassava baked hard like biscuit, and also some made out of a mixture of maize and yucca, which makes a good [kind of] bread" (1934, pp. 424–425). According to the earliest account of manioc processing, the roots were submerged for five days, after which they were removed, peeled, and crushed in a mortar. The juice was expelled by squeezing the pulp in a basketry sleeve. The resulting mass was then grated and sifted to produce flour, which was spread by hand on a griddle and compacted with a gourd to produce a large flat cake. Fermented drinks were made from cacao, manioc, and other ingredients and stored in large jars.

Among other kinds of food, turtles are most frequently mentioned. The Spaniards encountered large numbers almost everywhere they stopped. In one village, the corrals were estimated to contain six or seven thousand animals, and there were rarely less than 100. Since each turtle was "larger than a good sized wheel," one was sufficient to feed a family. A tasty and nourishing oil was extracted from the eggs. Acuña describes the methods of capture:

> They make large corrals, surrounded by posts, excavated inside so that they capture rainwater and become shallow lakes.
>
> This done, when the time comes that the turtles emerge to lay their eggs on the beaches, they also leave their houses to go to places where

the turtles are known to appear. There they wait until the turtles begin to dig the holes where they will lay their eggs. The Indians then cut off their retreat to the water, and fall upon them suddenly with the result that in a short time they have captured a large quantity with no more work than to turn them upside down so they are unable to move, providing all the time they want until they are all strung by holes made in the shell into various strings; thrown into the water they are towed behind the canoes without any difficulty until they can be placed in the waiting corrals, where they are kept imprisoned and fed with branches and leaves to keep them alive as long as desired (1942, p. 39).

Fishing was a major subsistence resource, since fish could be obtained daily in great abundance. The usual weapon was the spearthrower, but poisoning was also employed during low water. The favorite aquatic quarry, however, was the manatee, which was not only delicious but so nourishing according to Acuña that "with a small amount a person is more satisfied and more energetic than if he had eaten twice the amount of mutton" (1942, p. 38).

Little is said in the early accounts about terrestrial hunting or food gathering. Palm fruits were collected during the rainy season, however, and Brazil nuts were an important ingredient of the diet. Orellana's men found quantities of honey in some villages. Tapir and peccary were hunted on the terra firme.

OTHER ACTIVITIES. The principal weapon for hunting, fishing, and fighting was the spearthrower. It took the form of a flat board about 40 inches long and three fingers wide, with a bone hook at the upper end to secure the projectile. The spear or arrow was about six feet long ("nine palmos") and had a point of bone or very hard wood, which was sometimes detachable, permitting it to remain in the victim. To shoot, "the arrow is taken in the right hand, with which the spearthrower is held by its lower end, and placing the arrow against the hook, they launch it with such force and accuracy that they do not miss at fifty paces" (Acuña, 1942, pp. 51–52). Shields used in warfare were the height of a man, and either made of basketry or covered with cayman, manatee, or tapir hide.

In the absence of stone, axes and adzes were made from turtle shell. A piece about eight inches square was cut from the underside of the shell (which is the strongest part), cured by smoking, and then given an edge by grinding it on a stone. When inserted into a handle, an ax of this material would cut very well, although less rapidly than metal. The jaw of a manatee was used for hafting adzes. Planks, tables, seats, canoes, and other wooden objects were carved with these tools, and with chisels, gouges, and burins made of animal teeth inserted into wooden handles, which Acuña observed to be "no less efficient for the task than those made of fine steel" (1942, p. 54). Cedar trunks carried down from the highlands by high water were used for house timbers and making dugout canoes.

Cotton was spun and woven by the women, who produced not only all of the cloth required for local consumption but a surplus for trade to neighboring tribes, who were attracted by the fineness of the weaving. Very beautiful colored patterns were produced either during weaving or later by painting.

Pottery making was another highly developed art, to judge from Carvajal's description of ceramics encountered in one Omagua village:

> . . . there was a great deal of porcelain ware of various makes, both jars and pitchers, very large, with a capacity of more than twenty-five *arrobas* [one hundred gallons], and other small pieces such as plates and bowls and candelabra of this porcelain of the best that has ever been seen in the world, for that of Málaga is not its equal, because it [i.e. this porcelain which we found] is all glazed and embellished with all colors and so bright [are these colors] that they astonish, and, more than this, the drawings and paintings which they make on them are so accurately worked out that [one wonders how] with [only] natural skill they manufacture and decorate all these things . . . (1934, p. 201).

Other manufactured objects included baskets, trumpets, flutes, and drums.

While the provincial chief is the only occupational specialist specifically mentioned in the extant descriptions, the fine quality of textiles and pottery implies that at least part-time specialization existed in these crafts. The emergence of a full-time priestly profession is implicit in the varied responsibilities and high prestige of the religious practitioners, who served the people as "teachers, preachers, advisers, and mentors" (Acuña, 1942, p. 57). They were consulted to explain puzzling phenomena; they provided poisonous herbs for use in revenge, and they accompanied the warriors into battle, employing their skills to bring victory.

SOCIAL ORGANIZATION. Each village had a chief and all the villages in a "province" were united under a high chief, who was described by Carvajal as "a very great overlord and one having many people under him" (1934, p. 190). The Omagua ruler at the end of the seventeenth century was called Tururucari, which meant "god". His domain extended along the river for more than a hundred leagues, and he was obeyed universally "with great submission." The rulers of the provinces of Omagua and Machipero (which occupied the region to the east along the river) were friends and joined forces in warfare against inland tribes.

At the opposite end of the social scale from the chiefs were the slaves, who had been captured as children during raids on forest tribes. They were used for agricultural work and domestic tasks. Their role and treatment have been described by Fritz:

Every one has ordinarily in his house one or two slaves or servants of some tribe of the main-land, that he acquired in the course of war, or bartered in exchange for iron implements or clothing, or some other like way. The Omagua haughtily stretched in a hammock in lordly fashion despatches his servant or serving-maid, his slave or slave girl, to provide his food, bring his drink or other similar things. In other respects they regard their servants with much affection, as if they were their own children, provide them with clothing, eat from the same dish, and sleep with them beneath the same awning, without causing them the slightest annoyance (1922, pp. 48–49).

LIFE CYCLE. Newborn infants were buried alive if the mother was still nursing a previous child or if a girl was born when the parents wanted a boy. An inquiry by one of the missionaries elicited the information that many women had killed two or even three offspring.

Marriage involved the payment of bride price as well as five years of service by the groom to the father of the bride.

The death of an adult was believed to be caused by sorcery. The body was wrapped in cotton blankets and interred inside the house. The funeral rites, which lasted several days, featured continual lamentation interspersed with feasting and drinking.

CEREMONIES. The Omagua were "very much inclined" to festivals and dances, which lasted two to four days and involved the consumption of large amounts of beverages made from manioc, maize, or sweet potatoes. These gatherings provided the opportunity for holding conferences to plan raids for revenge and the capture of slaves.

TRADE. Cotton cloth was manufactured for trade with neighboring tribes.

WARFARE. The Omagua were in a continual state of war with tribes of the interior. For defense, their villages were stockaded or located on islands, where they were inaccessible to the canoeless mainlanders. The principal motivation for raids was revenge or the acquisition of slaves. Old men and women not suited for slavery were killed immediately, while captives of high status or outstanding courage, and who therefore constituted a potential danger if left alive, were put to death during ceremonies. The heads were kept in the houses as trophies.

Carvajal graphically describes the attack by the Omagua on the first Europeans to penetrate their territory:

> . . . we saw coming up the river a great many canoes, all equipped for fighting, gaily coloured, and [the men] with their shields on, which are made out of the shell-like skins of lizards and the hides of manatees and of tapirs, as tall as a man, because they cover him entirely. They were coming on with a great yell, playing on many drums and wooden trumpets, threatening us as if they were going to devour us (1934, p. 190).

RELIGION AND MAGIC. The Omagua believed in numerous spirits, which were represented by idols. Some had power over water, others over gardens, and still others over victory in war. The idol of the war god was carried in the prow of a canoe to insure the success of raids. Carvajal has described the construction and appearance of two idols in some detail:

> . . . in this house there were two idols woven out of feathers [or palm leaves] of divers sorts, . . . and they were of the stature of giants, and on their arms, stuck into the fleshy part, they had a pair of disks resembling candlestick sockets, and they also had the same thing on their calves close to the knees: their ears were bored through and very large, like those of the Indians of Cuzco, and [even] larger (1934, p. 201).

A special structure was erected for religious use. The idols were kept in it, as were the remains of deceased priests. It also served as a place where the priests could commune with the spirits. Captives were sacrificed to the idols and their heads were preserved in the shrines as religious trophies.

THE TAPAJOS

Location and Environment. The Tapajós is a clear water river that enters the lower Amazon approximately half way between the Rio Negro and the island of Marajó (Fig. 17). At this point, the main channel skirts the south edge of the várzea, causing partial blockage of the Tapajós mouth and the formation of a terra firme lake (Fig. 18). Fingers of the Guayana and Brazilian highlands attain their maximum southern and northern extensions in this region, creating a more varied topography than elsewhere along the margins of the flood plain. The changed appearance of the landscape impressed Carvajal, who described it as "the pleasantest and brightest land that we had seen and discovered anywhere along the river, because it was high land with hills and valleys" and "as normal in appearance as our Spain" (1934, pp. 223, 217).

This "normal appearance" probably refers to the large stretches of natural savanna, which are especially prominent along the left bank as a consequence of local climatic conditions. The only portion of the Amazon basin that receives an annual rainfall under 80 inches lies between 52° and 56° west latitude (Fig. 6). The precipitation is not only unusually low but it is also concentrated between December and June. During the intervening dry season, relative humidity may drop to less than 70 percent. When combined with well drained soil, this moisture is insufficient to sustain forest vegetation on the terra firme. The várzea, being dependent on the river fluctuation rather than on local rainfall, is less affected by these local climatic conditions.

At the time the Amazon was discovered, the Tapajós region was thickly populated. The sight of so many people apparently discouraged Orellana's crew, exhausted by the battles they had already been through. At any rate, Carvajal confesses that "so numerous were the settlements which came into sight and which we distinguished on the said islands that we were grieved" (1934, p. 218). Later, he adds somewhat testily that the villages "on account of their being [so] numerous could not be counted, and furthermore no attempt was made to do this [counting] because they did not leave us alone long enough for us to do so" (op. cit. p. 436). By the beginning of the eighteenth century, however, only a few scattered remnants of the indigenous population survived,

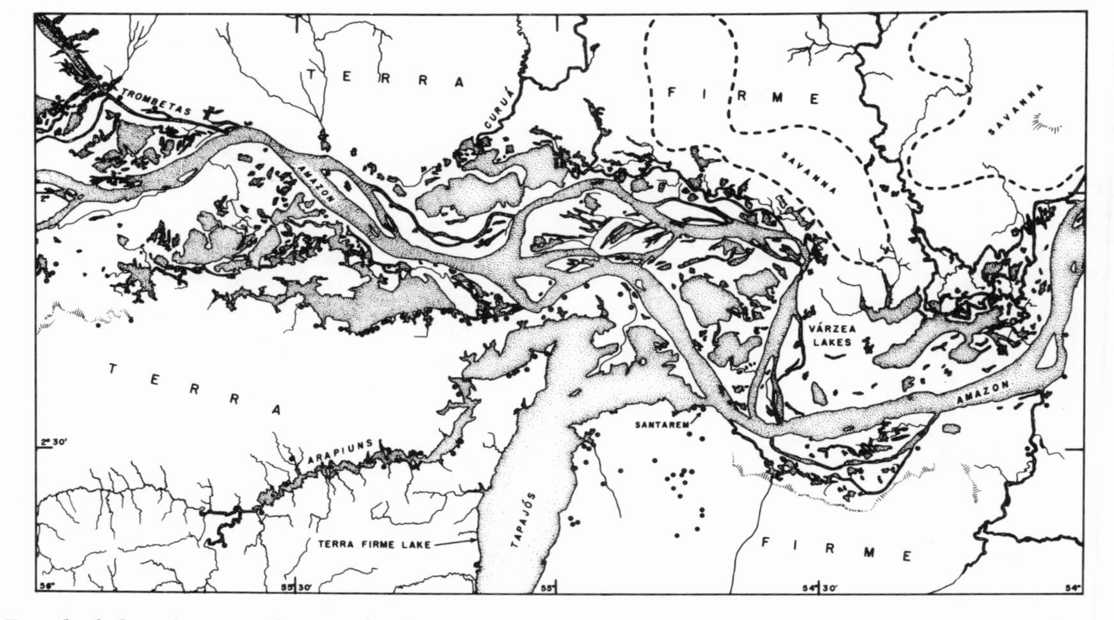

Fig. 18. Detail of the várzea at the mouth of the Rio Tapajós. The dense configuration of channels and lakes ceases abruptly at the edge of the terra firme. Partial blocking of the Tapajós mouth has created an extensive terra firme lake along the lower course of the river. Broken lines enclose patches of savanna. Dots represent archeological sites with Tapajós style pottery, few of which are located on the várzea. (Geographical detail after USAF Work Sheet 947B and 947C; archeological information from Nimuendajú, unpublished map on file at the Göteberg Ethnographical Museum.)

the majority having succumbed to slave raiding, missionization, disease, and other introductions of European civilization. Deculturation proceeded so rapidly that the linguistic affiliation of the Tapajós Indians is unknown, except that it was not Tupian.

Data on the precontact culture of the Tapajós are less detailed than for the Omagua, and other information from the general region has consequently been incorporated into the description. Carvajal's observations in 1542 are supplemented by those of Acuña in 1639. The principal documentary source, however, is Heriarte's history, written in 1692. Additional details are provided by archeological investigations.

SETTLEMENT PATTERN. Carvajal describes a settlement on the left bank of the Amazon that "stood in the bend of a small stream on a very large piece of flat ground more than four leagues long. This village was laid out all along one street and [had] a square half way down, with houses on the one side and on the other" (1934, p. 212). The inhabitants were sufficiently numerous that they "stirred up fear" in the European explorers. Population density was such that "one village was not half a league away from another, and still less than that along that whole bank of the river on the right, which is the south bank; and I can add that inland from the river, at a distance of two leagues, more or less, there could be seen some very large cities that glistened in white" (op. cit., pp. 216–217). Heriarte states that a typical Tapajós village contained between 20 and 30 houses and that the capital, which was located at the mouth of the Tapajós river, was the "largest town and population known so far from the district" and able to furnish 60,000 warriors. Acuña spent some time at a town of more than 500 families. An English expedition that explored the lower Amazon in 1628 also reported "many townes well inhabited, some with three hundred people, some with five, six or seven hundred" (Ashburn, 1947, p. 18).

These eyewitness reports of numerous large villages are substantiated by archeological evidence. In a survey made in 1923–1926, Nimuendajú located 65 sites, which he estimated to constitute less than half the number actually present in the area (Fig. 18). Except for a few fishing stations, all were above flood level and the majority were on hilltops. A considerable stability of residence is implied by the discoloration of the soil, which makes sites easily recognizable and gives them the local designation of "terra preta" or "black earth." Another indication of village permanency is the existence of straight sunken roads, about three to five feet wide and depressed about 12 inches, which connect the settlements. Wells were dug to supply water, since streams do not exist in the region. The village sites along the Rio Arapiuns are sufficiently large for the modern residents to utilize them profitably for commercial agriculture. Much of the modern city of Santarem overlies one of the most extensive of these "terras pretas," which has a thickness up to five feet. Shallower deposits occur along the lakes, suggesting shorter and perhaps seasonal occupation during the fishing season.

Dress and Ornament. Carvajal describes the Indians of the Tapajós as very tall with their hair cut short. They ordinarily wore no clothing, but for ceremonies had beautiful robes made by sewing feathers of different colors onto cotton cloth.

Subsistence. The principal crop was maize, which was grown on the várzea in large fields. Manioc and fruits were also cultivated. Maize was stored in baskets that were buried in ashes to protect the grain from weevils. Maize and manioc flour were mixed and baked into bread.

An important natural vegetable food throughout this part of the várzea was wild rice, which grew abundantly in the large lakes. It was made into bread and "very good wine." Various kinds of wild fruits were also eaten.

Although fishing was poor in the Rio Tapajós, fish, turtles, and manatees were obtained in large numbers from the várzea. Tapir, birds, and other game were hunted in the forest. "Turkeys," ducks, and parrots were kept alive in the villages.

Other Activities. Data on arts and crafts do not distinguish male from female activities. Pottery and wooden vessels employed for eating and gourds used for drinking were skillfully decorated with carefully drawn designs and painted in attractive colors. Carvajal makes special mention of the high quality of the pottery:

> They manufacture and fashion large pieces out of clay, with relief designs, [in the style] of Roman workmanship; and so it was that we saw many vessels, such as bowls and cups and other containers for drinking, and jars as tall as a man, which can hold thirty and forty and [even] fifty arrobas [about 4 gallons each], very beautiful and made out of a very fine quality of clay (1934, p. 442).

This description applies to the archeological ceramics from the area, which are characterized by thin walls, smooth cream-colored surfaces, and designs in low and high relief, as well as polychrome painting.

The principal weapon was the bow and arrow. Arrows were poisoned so that a mere prick of the skin brought death. Since no antidote was known, the Tapajós were feared by neighboring tribes and treated gingerly by the early Europeans.

Cotton was made into cord, which was woven into hammocks. Alternatively, palm fiber was used. Baskets were employed for storing maize and for many other purposes. Canoes were large enough to carry 20 to 40 people. Musical instruments included trumpets, drums, flutes, and rebecs with three strings.

A clue to the existence of craft specialization is Carvajal's remark that "the objects they manufacture would make a very good showing in the eyes of the highly accomplished artisans in that profession in Europe" (1934, p. 442), where advanced occupational specialization prevailed. Both the high chief and the religious leaders were probably also full-time specialists.

SOCIAL ORGANIZATION. Each village had a chief, and there was a high chief above them all, who was strictly obeyed by his subjects. Authority was determined by personal ability rather than heredity. Slavery existed but appears to have been less extensive than among the Omagua.

LIFE CYCLE. Polygyny was practiced and an adulturous wife was punished by drowning in the river. At death, a person was wrapped in a hammock and placed in a special structure with his possessions at his feet and a figurine at his head. After the flesh had disappeared, the bones were pulverized and mixed with wine, which was drunk by relatives and others.

CEREMONIES. Heriarte describes a festival held one evening each week in a cleared area behind the village. Trumpets and "sad and funereal" drums were played for an hour, after which singing, dancing, and wine-drinking began. The festivities ended when all the wine was consumed.

TRADE. Pottery and rice wine were traded to neighboring groups.

WARFARE. The Tapajós were greatly feared by neighboring tribes and maintained supremacy over them by virtue of their large numbers, poisoned arrows, and readiness to do battle. Female warriors were described by Carvajal "fighting in front of all the Indian men as women captains." They fought so fearlessly that "the Indian men did not dare to turn their backs, and anyone who did turn his back they killed with clubs right there before us" (1934, p. 214). The myth of the Amazons appears to be based on this observation.

Religion and Magic. According to several early authorities, painted idols were kept by the Tapajós in a special shrine, where they were presented with offerings of wine. This observation is substantiated by the occurrence of painted pottery figurines in the archeological sites. In addition, the bodies of dead chiefs were preserved and worshipped. Each person gave one-tenth of his maize crop to the gods; this grain was stored in the shrine and used to prepare the wine consumed in festivals.

Although religious practitioners are mentioned, the only duty attributed to them is the reading of omens.

SURVIVALS OF ABORIGINAL PATTERNS OF RESOURCE USE

Early postcontact information on the subsistence pattern of the aboriginal várzea dwellers can be supplemented by descriptions of recent hunting, fishing, and gathering activities. Although some of these are now conducted for commercial ends, the nature of the techniques employed suggests that they are survivals from pre-Columbian times.

Birds attracted in droves by the plentiful food supply in the shrinking várzea lakes are hunted by the modern inhabitants at night. They enter the feeding grounds quietly, lighting the way with a faint kerosene lamp carried on the leader's head in a small box open only at the front. A man on either side of the leader carries a net or pole. At the proper moment, the net is cast and the pole is wielded horizontally, breaking the necks of the trapped birds. The rest of the flock is frightened off but returns if the hunters remain quiet, so that the operation can be repeated. In the vicinity of Manaus, thousands of ducks are killed annually by this method.

Early nineteenth century explorers observed the Indians of the lower Rio Japurá preserving large quantities of birds captured in this way by roasting them over an open fire and then packing them between palm fronds. These bundles were stored under the ridgepole of the house.

Turtles are still taken by the hundreds during low water, when they crawl onto the beaches to lay their eggs. Adults are kept alive in corrals at the water's edge until needed. The eggs are sometimes preserved by drying in the sun or over a fire, with the result that they lose one-third of their weight and take on a "repugnant greasy flavor." Large numbers are processed into "butter" in dugout canoes, which have been dragged onto the beach where the fresh eggs are collected. After the sand has been washed off, the eggs are put into the dugouts with a little water and trampled until completely mashed. Three days' exposure to the sun brings the oil to the surface, where it is ladled off with wooden spoons into large pottery jars. Manatee and turtle meat is preserved in this oil after being cut into small pieces and fried. Alternatively, manatee oil is used for this purpose. Newly hatched turtles cooked in tucupí (manioc juice and pepper) are a special delicacy.

During August and September, when the water level is dropping, the modern inhabitants of the middle Amazon abandon their houses for temporary fishing camps at the edge of the várzea lakes, two or three days distant. This seasonal migration is a festive event:

> It is impossible to imagine the joy with which these people move. No more genuine fraternity could prevail, especially when they gather in the shade of the trees to prepare their meals. They joke, laugh, play the guitar, and sing. Continuing the voyage, they enter the lake and chose a site on the margin, which may have been used on previous occasions, to construct their thatched huts and the shed with pole racks where the slabs of salted pirarucu will be laid to dry. After two months the harvest is over, provisions are exhausted, and the rains are beginning. The fishermen load their produce in the boats . . . , break camp together, and everyone returns home well satisfied. . . . The following year at the same time, another pilgrimage and another abandonment of homes for the fishing lakes (Bitencourt, 1951, 138–139).

Although today a large part of the catch is sold, it is likely that a similar pattern of intensive seasonal exploitation existed during the aboriginal period, when the surplus would have supplied the needs of the village during the rainy season.

In addition to wild rice, the *Victoria regia* is abundant in the várzea lakes. The dark rounded seeds, which are produced in a spherical capsule the size of a child's head, resemble large beans. They can be eaten either roasted or ground into flour and made into cakes. The tuber is also edible.

In the vicinity of Santarem, minute shrimp captured in nets and dried in the sun are considered a delectable tidbit. Cayman eggs are eaten, in addition to the reptile itself. Fat extracted from the flesh is used to prepare ungents and paints. Although the dolphin is edible, it is tough and has an oily taste; a century ago it was still hunted, but superstition causes it to be avoided by the modern residents.

The tambaqui, one of the largest and most abundant of the Amazonian fishes, is easily captured when the water is low by capitalizing on its novel habits. It subsists on fruits, which it locates by sound when they drop into the water. The natives imitate this impact with a small stone attached to one line, while another line is baited with the fruit. During low water, this fish is caught in such quantity that it finds no buyers in the Manaus marketplace.

A nonsubsistence product of the forest is the silky white fiber of the sumaumeira or silk-cotton tree; a century ago it was still being used by the Indians to make beautiful fabrics.

CHARACTERISTICS OF THE VARZEA ADAPTATION

Although the aboriginal cultural pattern of the várzea shares many features with that of the terra firme, it also differs in significant ways. The similarities are those traits that make the várzea complexes an integral part of the tropical forest culture area. These similarities include the staple cultigens; villages composed of communal houses occupied by an extended family; and manufactures, such as hammocks, mats and baskets, pottery, and feather ornaments. The differences manifest themselves in weapons, in social and political organization, and religious practices, many of which have close parallels in the Andean area, from which some of them were undoubtedly derived. On the other hand, the adoption of these traits, most of which represent advances over the level of complexity achieved in the terra firme environment, was made possible by the more propitious environment of the várzea.

The principal cultural differences and their adaptive significance can best be discussed by employing the same frame of reference utilized for the terra firme situation: namely, techniques for maximizing food return and techniques for controlling population size and concentration.

Techniques for Maximizing Food Return

Although the principal defects of the terra firme environment from the standpoint of subsistence exploitation—rapidly declining soil fertility and low concentration of protein resources—are absent from the várzea, other limitations on agricultural productivity exist. Since they are of different kinds, however, different forms of technology and sociopolitical and religious behavior might be expected to develop during adaptation. Several features of Omagua and Tapajós culture appear to reflect this situation.

The dominant factor on the várzea is the regime of the river, which regulates the annual cycle of plant and animal life and consequently the subsistence opportunities available to man. Low water is not only a time of concentrated abundance (even superabundance) of wild foods, but also of agricultural activity (Fig. 19), while the period of high water is characterized by a relative scarcity of wild plants and the dispersed distribution of aquatic fauna. Although fishing and hunting are possible during high water, the return per man hour of labor expended probably drops close to that of the terra firme habitat. The foremost adaptive problem is thus the prolongation of plenty into the time of scarcity. This can be done in two principal ways: 1) by preserving and storing plant and animal foods for later use; and 2) by developing an occupational division of labor in subsistence activities, so that simultaneously available foods can be intensively exploited.

The early accounts provide abundant evidence of food preservation and storage by the várzea inhabitants. The following statement summarizes the situation:

> They are people of considerable foresight, and they keep provisions on hand until the time when they take in the next harvest, and they have others in store in lofts or on hurdles raised above the ground the height of a man and [sic] as high as they see fit; and they keep their maize

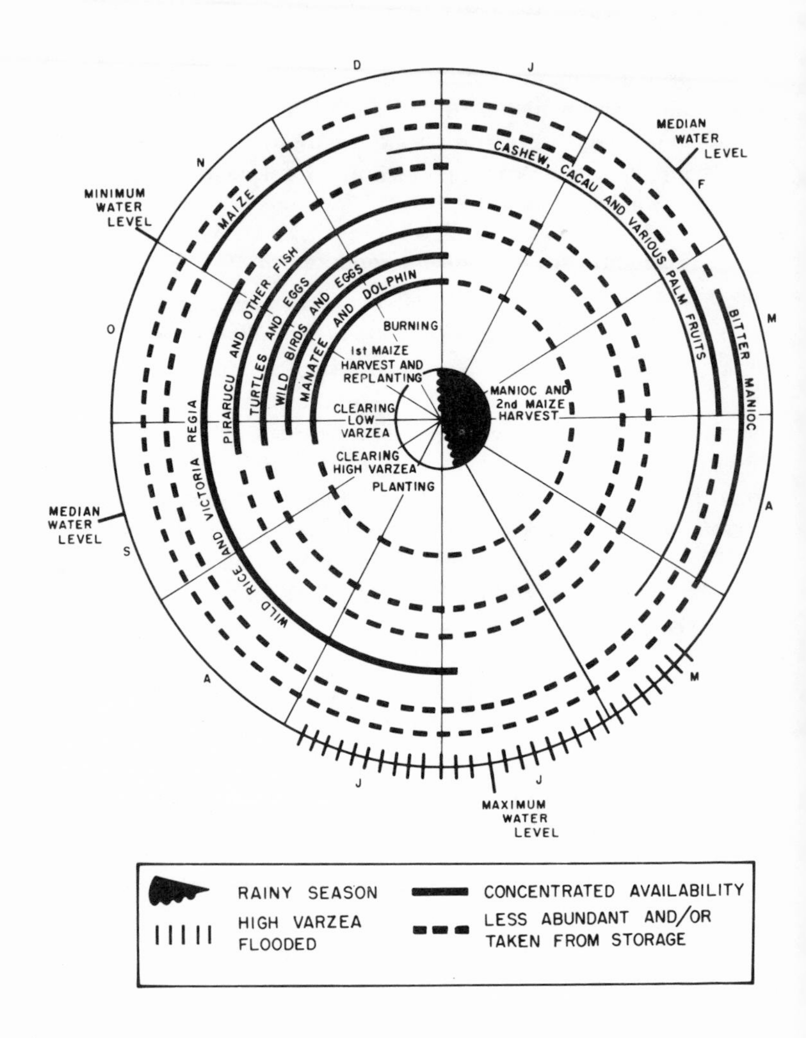

Fig. 19. Annual subsistence round postulated for the aboriginal inhabitants of the várzea. Wild plant and animal foods are available in profusion during the months of October through December, when the water level is receding. Maize and bitter manioc, the principal agricultural crops, must be harvested before the fields are flooded. The fluctuation between feast and famine was mitigated by the development of a variety of methods for preserving and storing seasonally available foods.

and their biscuit (which they make out of maize and cassava mixed together or bound with a paste), and a great deal of roasted fish, and many manatees, and game meat (Carvajal, 1934, p. 398).

Turtles, captured alive and kept by the hundreds in pens in the villages, served as a convenient source of fresh meat during the flood period. Meat and fish were also preserved in jars filled with manatee or turtle egg oil, or by drying and smoking. Maize was stored underground in baskets by the Tapajós. The Omagua, who subsisted principally on bitter manioc, dug up the tubers and reburied them in pits where they could be removed as needed. Wild rice, which was difficult to store under humid conditions, was converted into wine.

While occupational division of labor in subsistence activities is not mentioned by any of the early travelers, its existence can be inferred both from evidence that other types of full-time specialists had developed among várzea groups and from the fluctuating pattern of food resource availability. In contrast to the terra firme, where an annual subsistence round can be constructed to take advantage of sequentially available wild and domesticated foods, the várzea alternates between abundance and scarcity. During the period of low water, all foods are simultaneously at peak abundance and must be gathered in sufficient quantity not only to satisfy immediate daily needs but to accumulate a surplus for consumption during the months of reduced productivity. The most efficient way to cope with such a situation is to split the labor force and allocate to each group a specific type of activity. Furthermore, agricultural operations on the várzea must be systematically organized because of the time limitations imposed by the regime of the river. If planting is delayed too long, the crop will not mature before inundation; if planting is done too soon, seeds or cuttings may rot before they sprout. Timing is even more critical when two harvests are programmed, as they were in the case of maize. The expertise needed to manage várzea agriculture efficiently would almost inevitably have led to the emergence of specialists, who could direct part of the work force while other members of the community concentrated on hunting, fishing, and gathering activities. Some form of labor allocation is not only a logical response to the pattern of seasonal abundance, but is also compatible with the degree of social stratification evidenced by the

Omagua and the Tapajós. Both were ruled by a high chief who had the power not only to issue orders but to command obedience. Furthermore, the fact that incipient division of labor in subsistence activities occurs among some terra firme tribes suggests that adaptation to the várzea situation could have evolved merely by intensifying characteristics already possessed by one or more local groups.

Although várzea subsistence resources were inexhaustible under aboriginal methods of exploitation and management, they were subject to erratic fluctuations in abundance that injected a degree of insecurity into the lives of those dependent on them. The danger lay in the unpredictability of the annual flood crest, which at irregular intervals rose to inundate land usually free of water, or remained abnormally low, withholding fertile silt from the fields. Since the flow of the Amazon is determined by rainfall patterns hundreds of miles away, its height could not be foreseen by people living along the banks. Furthermore, even if prediction were possible, the volume of water ruled out practical measures of control. Elaboration of religion is a characteristic response to subsistence insecurity and it is consequently not surprising to find the várzea inhabitants described as "idolaters." Images of pottery, wood, or basketry were kept in shrines, where they were attended by priests; and offerings were brought to them by worshippers. Some of the deities had special powers over cultivated plants, while others were influential in warfare and other kinds of activities. Concomitantly, the rather limited role of the shaman was elaborated into that of a priest, who was probably a full-time religious practitioner.

Techniques for Control of Population Size

The aboriginal settlement density on the várzea differed markedly from that of the terra firme. The banks of the river and many of the islands were inhabited, and in some cases an unbroken succession of houses extended for long distances. The village plan was linear, either along the shore or on both sides of a street, in contrast to the circular arrangement characteristic on the terra firme. Residence stability also was greater. Village population size, however, appears not to have differed significantly from the upper part of the terra firme range. Although one Tapajós settlement is reported to have had a population of 2,500, 500 to 700 inhabitants seems to have been more typical and Omagua villages averaged somewhat smaller.

While the similarities between the terra firme and várzea in village size and extended family household composition reflect the basic uniformity of Amazonian culture, the differences represent adaptations to the distinctive aspects of the várzea habitat. A linear arrangement of houses is natural for a settlement oriented toward exploitation of the várzea and is the characteristic pattern for river bank settlements everywhere. Since access to the várzea resources is greatest along its margin, settlements will expand laterally before they begin to spread inland. One would expect on this basis that an equilibrium adaptation would take the form of a narrow band of almost continuous settlement along the margins of the floodplain, and this is exactly what was reported by the first European visitors.

Although várzea resources were not exhaustible under aboriginal methods of exploitation, they were subject to temporary depletion from natural causes. An equilibrium adaptation should consequently stabilize the population at a size that could be supported during lean years rather than at the much higher level permitted by optimum or even normal conditions. A population that was allowed to increase when resources were abundant would be vulnerable when they were suddenly and drastically reduced. The resultant violent fluctuations in population size would endanger the stability of the cultural configuration essential to the survival of the community. In this context, possession by the várzea peoples of several kinds of cultural traits that offset population increase among terra firme groups becomes intelligible. While prenatal controls are not mentioned, infanticide was practiced frequently by the Omagua. The Tapajós punished adultery by females with death (as did the Jívaro). In the case of the Jívaro, this penalty is correlated with sexual restrictions on males (as we saw in a previous chapter), suggesting that similar taboos may have been observed by the Tapajós. Blood revenge by poison and murder was practiced by the Omagua, and both várzea groups fought constantly with their terra firme neighbors. The Omagua killed certain types of prisoners, including those of advanced age, high status, or outstanding courage. The existence of such customs implies a strong necessity to offset natural tendencies toward population increase in spite of the superficial appearance of várzea subsistence abundance.

In the case of the Omagua, this proposition seems at first glance to be negated by the practice of taking captives, which makes possible a rate of population growth more rapid than normal reproduction. This apparent discrepancy can be resolved by comparing the position of captives in Omagua society with that among terra firme groups. A sharp distinction immediately becomes evident: whereas terra firme captives are adopted into the kinship system and thereby assume the same rights and privileges as members born into the community, Omagua prisoners were viewed as property and their activities were defined by their servant status rather than by kinship relations. Although they augmented the labor force, there is nothing to indicate that they possessed any skills not represented among the Omagua population. In other words, they were not an essential element in the social system, but rather an appendage that could be eliminated without seriously affecting its normal operation. This situation suggests the possibility that the function of slaves in várzea society may have been to serve as a kind of buffer between the social organism and the environment. In good years, their presence would have permitted more complete utilization of the available subsistence resources. When food was scarce, however, they were expendable without danger to the integrity of the society as a whole. Similarly drastic fluctuations in the size of the Omagua population, on the other hand, would have been intolerable because they would have removed too many links from the mesh of reciprocal kinship obligations on which social integration and therefore survival depended.

Techniques for Control of Population Density

Although one of the most conspicuous traits of the várzea inhabitants was their bellicosity, the density of the population reported by the first observers indicates that warfare did not operate as a spacing mechanism as it did in the terra firme. Significantly also, the primary focus of aggression appears to have been between várzea groups and their terra firme neighbors; raiding across várzea frontiers appears to have been minimal, although defensive measures existed in the form of forts and unoccupied "buffer" zones. The Omagua chief is even reported to have contracted an alliance with the chief of the adjacent province for mutual defense against terra firme neighbors.

Extant information does not indicate that sorcery or other kinds of behavior serving to inhibit population concentration on the terra firme functioned similarly on the várzea, and it seems probable that differences in subsistence availability made such sanctions inappropriate. The susceptibility of most of the várzea to annual inundation, and of all of it to intermittent submersion, reduced its suitability for permanent settlement. On the other hand, the width of the productive area was sufficiently great that a nearly continuous population distributed along the margins was not only compatible with indefinite subsistence exploitation, but probably was the most efficient type of settlement pattern because it placed a maximum amount of manpower in the position of greatest accessibility, minimizing travel time to fields and fishing grounds and the burden of transporting produce to the village. Under such conditions, behavior preventing the development of contiguous or at least closely spaced villages would have no ecological justification.

If warfare has an adaptive function, the low level of hostility between várzea populations can be interpreted as an indication that little benefit would accrue from its intensification. Several other considerations support this hypothesis. Because unpredictable devastation is an equal threat over the entire várzea, subsistence security cannot be enhanced by territorial expansion along the river. Nor would any mutual gain result from warfare between várzea provinces for the purpose of taking each other captive. On the other hand, conflicts with residents of the adjacent terra firme were facilitated both by an extensive common frontier and by the concealment provided by the forest cover. More important, such warfare performed functions of a reciprocal and mutually beneficial nature, helping to curb population increase among terra firme groups and supplying the várzea societies with captives, which provided the "safety valve" needed for optimum exploitation of floodplain subsistence resources.

If the foregoing analysis is generally correct, it indicates that the potentiality of the Amazonian floodplain for the development of civilization is not comparable to that of the fertile river valleys of Asia and the Near East. The general parallels in population density, level of cultural complexity, and pattern of hostility between floodplain and inland peoples that evolved in all these regions have diverted attention from the existence of a fundamental distinction in the pattern of resource availability. In the more temperate river basins, seasonal differences in food supply are felt least by the floodplain farmers, who have a storable harvest to draw upon, and most severely by hunters and gatherers in the interior, who are unable to make adequate preparation for winter scarcity. In Amazonia, in contrast, overall subsistence security is greater on the terra firme, because of the year-round productivity of the staple foods, than on the várzea, where six months of superabundance alternate with six months of relative privation. The environmental imperatives ruled out the possibility of intensification of both patterns, and in so doing eliminated Amazonia as a potential cradle of higher civilization.

Andean Influences on the Culture of the Várzea

The higher level of cultural development achieved on the várzea leads us to ask whether it is mainly a reflection of the greater accessibility of this portion of the Amazonian lowlands to influences emanating from the Andean area, or whether it can be explained as a consequence of local evolution. A number of the cultural traits possessed by the Omagua are clearly derived from the highlands. For example, they were the only lowland group reported to practice skull deformation at the time of European contact, and one of the few to use shields for defense in warfare, both of which are ancient Andean traits. Their principal weapon, the spear and spearthrower, is also characteristic of the Andean area. Finally, they wore cotton clothing resembling that of highland peoples, although it is poorly adapted to conditions of humid heat.

The Omagua and Tapajós also manifest a number of other traits of a sociopolitical and religious nature that have counterparts in the Andean area, where they emerged during the early Formative period, about a millennium before the beginning of the Christian era. Among these are temples, religious articles (including idols), political organization joining several settlements under a permanent leader, occupational division of labor, and social stratification (Table 5). Is it probable that these traits were also intrusions into the várzea from the Andean area?

Table 5. Cultural traits of evolutionary significance among várzea groups. (X? = Probably present.)

Traits	Omagua	Tapajós
SEDENTISM		
Village population	300+	300–2,500
Village permanency	Indefinite	Indefinite
SOCIAL ORGANIZATION		
Household chief	X	X
Multihousehold chief		
Village	X	X
Multivillage	X	X
SOCIAL STRATIFICATION		
High Chief	X	X
Slaves	X	X
FULL-TIME OCCUPATIONAL SPECIALISTS		
High chief	X	X
Shaman	X?	X?
Arts and crafts		X?
TRADE		
Within village	X?	X?
Between villages or intertribal	X	X
Formal market		
SPECIALIZED STRUCTURES		
Temple or shrine	X	X
Storehouse or granary	X	X
Chief's house	X?	X?
RELIGION		
Idols	X	X
Prayers and offerings		X

Before attempting to reply to this question, it is necessary to make a distinction between two general categories of cultural traits. One category is represented by the weapons, clothing style, and head flattening borrowed by the Omagua from their Andean neighbors. In substituting these for their lowland equivalents (bow and arrow, nudity, earlobe or lip perforation), the Omagua did not significantly affect their adaptive relationship to their environment. Although clothing is disadvantageous in humid heat, the men at least are reported to have removed theirs during periods of physical activity. Under várzea conditions, the spearthrower is probably as efficient as the bow. In short, replacements of this kind can be made with relative ease and without significantly affecting the basic structure of the society.

The distinctive Omagua and Tapajós sociopolitical and religious features, on the other hand, are not simple substitutions for previously existing traits. In several cases, they are intensifications or elaborations of tendencies that are vaguely manifested by a few tribes of the terra firme. The Kayapó and Camayurá, for example, had permanent village chiefs and buildings for other than household use, while occupational specialization in arts and crafts and physical representation of spirits occur in incipient form among the Camayurá (Table 4). The appearance of these features can be attributed to the higher population concentration achieved by these two tribes, and it follows that their further elaboration by the Omagua and the Tapajós resulted from the even greater population density attained on the várzea. If this correlation between cultural complexity and population concentration is valid, it is obvious that no amount of familiarity with these practices in another society could provoke their adoption before a certain threshold of population density was reached. It also follows that once this critical density has been attained, effective adaptation not only permits but requires the development of new integrative mechanisms, so that these will emerge whether or not a model is available.

The existence of an intimate relationship between environmental potential and level of cultural development also implies that a group forced to move into an area with reduced resources will be unable to maintain its previous level of development if that level is unsuitably complex. An archeological example of the simplification that accompanies adaptation to a lower subsistence productivity is provided by the prehistoric Marajoara culture on the island of Marajó at the mouth of the Amazon. When this culture appeared on the island, it seems to have possessed a more highly stratified society than that of more recent várzea groups, such as the Omagua and Tapajós. During its history on Marajó, it underwent a decline in complexity that is reflected archeologically in the disappearance of elaborate kinds of pottery decoration and special ritual practices and articles.

Conclusion

Although information on the aboriginal cultural adaptation to the várzea habitat is fragmentary, it clearly indicates that population concentration was greater and the level of sociopolitical complexity more advanced than on the adjacent terra firme. This was not the result of an improvement of the subsistence resources by man, but rather of a sensitive cultural adaptation that permitted efficient utilization of the unique natural productivity of the várzea. This natural productivity has two important defects, however: 1) it is highly seasonal; and 2) it is subject to unpredictable fluctuation. Cultural mechanisms were developed to compensate for seasonality, but the food shortages that resulted from premature or prolonged inundation could not be predicted or offset. Adaptation to this situation set a ceiling on cultural development and it seems probable that várzea groups, such as the Omagua and Tapajós, had achieved the maximum level of cultural elaboration consistent with these local environmental conditions.

Chapter 5

AMAZONIA IN THE MODERN WORLD

During its long geological history, Amazonia has passed through a succession of distinct phases, each initiated by a major geological event that opened new ecological niches and closed old ones, forcing readaptation by the plant and animal life.

Man's advent a few thousand years ago probably made no significant impact on the equilibrium of the ecosystem because his numbers were initially few and his adaptive niche was broad. Before the human population increased to a size that might have been detrimental, natural selection had brought about a finely balanced adaptation to the environmental resources. As a result, aboriginal man appears to have been no more destructive than his fellow organisms to the long-term stability of the tropical rain forest ecosystem.

The arrival of European explorers at the beginning of the sixteenth century had very different consequences, for two reasons: 1) the primary aim was commercial exploration rather than settlement; and 2) close contact was maintained with the homeland, which dictated the kind of commodities to be supplied and their price. For the first time in its long history, Amazonia thus came under the continuing influence of an agent that was extra-continental and consequently immune to the molding forces of local natural selection.

During the first century following discovery, the disruption was minimal. After about A.D. 1615, however, French, Dutch, English, and Portuguese settlements began to proliferate around the mouth of the Amazon and political rivalry became increasingly intense. Each nation recruited thousands of Indians to defend its claim to sovereignty, and mortality was so great that by 1631, when the Portuguese emerged victorious, there were few natives left to exploit. In 1664, slave raids were being conducted as far west as the mouth of the Rio Negro; 30 years later only a few dozen captives could be obtained even along the upper Amazon, where thousands of Indians had lived a century before.

European activities might have been less devastating to the indigenous population if the colonists had not inadvertently introduced several virulent and highly communicable diseases. Before A.D. 1500, explorers reported Amazonia to be a remarkably attractive place. Sir Walter Raleigh (1811, pp. 153–154), who had traveled widely, asserted that

> For health, good ayre, pleasure, and riches I am resolved it cannot be equalled by any region either in the East or West. Moreover the countrey is so healthfull, as of an hundred persons and more . . . we lost not any one, nor had one ill disposed to my knowledge, nor found any Calentura, or other of those pestilent diseases which dwell in all hot regions, and so neere to the Equinoctiall line.

The intensified immigration during the seventeenth century brought this blissful situation to an abrupt end. A smallpox epidemic swept the lower Amazon in 1621, and another devastated the upper portion in 1651. Since the Indians had no immunity, often whole villages were wiped out. To make matters worse, the importation of African slaves led to the introduction of malaria and yellow fever. Nothing could be more diametrically opposed to the situation described by Raleigh than the condition of the population in 1913:

> Generally speaking, the inhabitants living upon the river banks show evidence of either acute or chronic disease or the effects of having suffered from such disease. Portions of Amazonia today constitute some of the most unhealthy and most dangerous regions to reside in, from the standpoint of health, that exist in the tropics (Stong and Shattuck, 1913; quoted in Ashburn, 1947, p. 115).

The decimation of the indigenous inhabitants, tragic as this was from a humanitarian standpoint, did not significantly affect the ecosystem as a whole. What did affect it was the replacement of aboriginal cultural practices with attitudes and behavior developed in a totally different environmental context and incompatible with local ecological conditions. Although the early explorers praised the healthy climate and marveled at the luxuriant vegetation, they were horrified at the head-hunting and infanticide practiced by the inhabitants. The native custom of frequently moving villages conflicted with the European system of private land ownership and permanent settlement. A scattered population could not provide the concentrated labor force necessary for plantation agriculture and for the production of export goods. Some of these "primitive" customs were eliminated (often forcibly) by the colonists and missionaries, while others were weakened through acculturation. The frailties of the modern cultural configuration of Amazonia are the inevitable consequence of this unnatural intrusion of an alien cultural pattern into the tropical forest environment.

One of the most striking characteristics of life in Amazonia today is the absence of regional differentiation. Along all the main rivers and many of the smaller tributaries, people eat the same food, wear similar clothing, live in the same kind of house, and share the same beliefs and aspirations. Having lost the ability to satisfy their needs from the resources of the forest, they are obliged to purchase not only cloth, pots and pans, knives and guns, but also many basic subsistence items, such as sugar, salt, rice, beans, and coffee. Payment must be made in terms of what the export market demands rather than what the local area might best produce. The principal activities are consequently rubber-collecting, gathering wild fruits and nuts, fishing, hunting for hides (especially jaguar, cayman, and peccary), agriculture, and cattle raising. The distance from the market, the high markup by the middleman, and a commercial organization that prevents the seller from seeking the best price combine to curtail the producer's profit. Illness, bad luck, unfavorable weather, or other extenuating circumstances often reduce a man's productivity below the minimum necessary to supply his family's needs. He is then forced to obtain credit from the local trader, and once this step is taken he becomes a slave to the system with no real hope of extrication. Unable to purchase adequate food or to spare the time for fishing and gardening, he and his children tend to suffer from nutritional deficiencies, which also lower their resistance to other kinds of disease.

The habitat has suffered a similar degradation. Overexploitation of the land surrounding large settlements has produced marked and probably irreversible deterioration of the soil and vegetation, causing local extinction of many species of animals and birds. Overexploitation and other types of disturbance have reduced the incredible precontact density of river turtles, caymans, water birds, and other forms of aquatic life to remnants that seek concealment in remote or inaccessible places. The whimsical tastes of purchasers half-a-world away for exotic pets, spectacular feathers, curious ornaments, or unusual foods have encouraged the decimation of plants, animals, and even insects. As scarcity increases, prices rise and efforts to satisfy the demand are intensified. Until recently, the small size of the acculturated population and the high cost of transportation have kept ecological damage confined to the várzea and the margins of the major tributaries. In the vast interior regions, which have either remained unoccupied or support remnants of aboriginal groups, little or no harm has yet been done.

Now, however, this reservoir is also threatened by spectacular population growth and intensified development programs of national and international inspiration. In 1940, the Amazon basin had 1,876,025 inhabitants. By 1950, as a result of the eradication of malaria, yellow fever, and other tropical diseases, the total had risen to 2,372,508 and by 1960 it was 3,569,066. At the present rate of increase, there will be over seven million people there by 1985 and more than 11 million in A.D. 2000. In view of the magnitude of the degradation that has occurred under a population density that is one-tenth of the latter figure, this prospect is appalling.

The projected population growth would be a sufficient basis for pessimism even if it stemmed from natural increase by the existing inhabitants, who have some familiarity with the problems of the environment. A large increment, however, comes from other parts of Brazil, particularly the arid northeast, where the myth of an Amazonian paradise continues to flourish. Colonists are actively encouraged by federal and state governments, which see population growth as the key to economic prosperity. Homestead programs similar to those that pushed the United States frontier westward are facilitating settlement along highways that slice through major tracts of virgin forest. The Belem-Brasília road, completed in 1960, and the newer Brasília-Acre road pass through previously inaccessible regions on the eastern and southern margins of Amazonia, while the Transamazonica route cuts across the center of the area. Its construction, which commenced in 1970, is accompanied by a vigorous plan of colonization fomented by the federal government, which offers social, technical, and financial aid to prospective settlers. This large-scale program is being expedited without regard to scientific reports published by the Instituto Brasileiro de Geografia, another federal agency, which assert that the area in unsuitable for intensive subsistence exploitation, and conclude that "a disaster of enormous proportions" is imminent.

Amazonia has also attracted the attention of armchair development planners in the United States. The most ambitious scheme is that proposed in 1967 by the Hudson Institute for construction of an earth dam across the lower Amazon, which would transform the várzea between Santarem and Tefé into an enormous lake. Among the benefits anticipated from this engineering feat are improved access to high land, facilitation of deep draft shipping, and stabilization of the delta, making it more amenable to agricultural utilization. Anyone who has digested the data in the foregoing pages will immediately recognize that such a scheme would wipe out the most potentially useful portion of Amazonia. The várzea above the dam would be permanently inundated; downstream it would be deprived of the annual sedimentation crucial to the maintenance of soil fertility. Fortunately, the adoption of this recommendation is unlikely—not because it is ecologically unsound, but because it is politically distasteful.

Ironically, the choice need not lie between irreversible devastation and no exploitation at all. If there is one lesson to be learned from ecology, it is that adaptation is profitable. The aboriginal inhabitants unconsciously developed an optimum utilization of the habitat under the slow manipulation of natural selection. Theoretically, we should be able to do better because we possess a body of scientific knowledge that we can apply consciously and intensively to the problem of increasing the natural productivity of the area. Until now, however, we have been devoting our greatest efforts to the ecological equivalent of forcing square pegs into round holes. For example, we attempt to convert the forest into grassland for raising cattle (which are ill-suited to tropical heat and humidity and derive inadequate nutrition from tropical vegetation), instead of using the natural pasture of the várzea lakes for the intensive breeding of manatees (which are well adapted to local conditions and produce a higher yield of meat and hides).

Another likely candidate for controlled exploitation is the large water turtle, now threatened with extinction. A recent study of the life cycle of this animal has shown that between 25 and 80 percent of its embryos are destroyed each year by inundation of the nests before the eggs hatch. In addition, predators consume 10 percent of the newly born turtles before they reach the safety of the water. Systematic salvage of these millions of individuals would not only provide food and income for many more people than the present opportunistic methods support, but would do so on a permanent basis. Fish culture is another industry of potentially high productivity and commercial profit. Silviculture, which mimics the natural forest vegetation, could be expanded on the terra firme.

Research will not solve the problem, however, while customs and legal procedures incompatible with the realities of Amazonia continue to prevail. The greatest asset of the várzea is the annual rise and fall of the river, which perpetually renews the fertility of soil and water. In the process, islands are dissolved and reformed, channels constantly change, and all landmarks are ephemeral. A system of private landownership is totally unsuited to this fluid situation. Faced with the prospect of unpredictable boundary changes, an owner is reluctant to invest in any improvements and is encouraged instead to extract the maximum profit while the opportunity lasts. Clearly, the solution is to return to the aboriginal system of cooperative exploitation, which was not dependent upon fixed property lines and which distributed profit and loss equitably. The alternative (namely, the attempt to canalize or dam the river to force it to behave in accord with alien social and legal requirements) would not only be enormously expensive, but would also have a negative effect on the natural productivity of the várzea.

As we approach the final quarter of the twentieth century, the fate of Amazonia hangs in the balance. Whether the next few decades will bring ruin or salvation is not yet predictable. Although the gaps in our knowledge are vast, we have enough information at our disposal to prevent irreversible degradation. On the other hand, the relatively high resiliency of temperate ecosystems encourages the delusion that the tropical lowlands can be reshaped with equal impunity, and the slowness with which even major disturbances set in motion ecological chain reactions makes their potential consequences easy to ignore. As long as the impetus for Amazonian exploitation comes from alien cultural roots, the possibility of a rational program of development is nil.

This prospect is the more tragic because *Homo sapiens* alone of living things has the capacity to view his surroundings on a more comprehensive scale than the span of a single life. When we assess man's activities in the perspective of geological time, we are forced to recognize that what is happening in the biosphere today is not commonplace. In fact, not since primordial organisms evolved the capacity to release free oxygen thousands of millions of years ago has any new species developed the capacity to alter the adaptive conditions for life on earth. Continents have changed in configuration, glaciers have advanced and retreated, seas have been elevated and mountains submerged, and the poles have moved around, but the physical and chemical parameters have remained essentially the same. Now, suddenly, new chemical compounds and old ones in abnormal concentrations are pouring into the water, land, and air. Just as the aboriginal inhabitants of Amazonia were nearly exterminated by Old World diseases to which they had no immunity, so plants and animals evolved over hundreds of millions of years are unable to cope with alien chemicals suddenly introduced into their habitat. Knowing in general how natural selection operates, we can confidently predict that a few of the millions of extant species will be "preadapted" to the new conditions, but there is no assurance that *Homo sapiens* will be among these survivors. Bringing mankind through this ecological crisis is the greatest challenge that culture as an adaptive mechanism has ever faced.

Chapter 6

THE EVOLUTIONARY SIGNIFICANCE OF ADAPTATION

Amazonia as it exists today is the product of millions of years of geological and biological evolution. The luxuriant vegetation that blankets its surface has made a remarkable adaptation to constantly warm temperature, high humidity, and soils divested of soluble nutrients. All of its distinctive characteristics, which include extraordinary species diversity, low frequency and scattered distribution of individuals of the same species, colossal nutrient storage capacity, evergreen foliage, and preponderantly vegetative method of reproduction, are adjustments to unfavorable climatic and edaphic conditions. The terrestrial fauna that subsists on this vegetation is similarly diversified and nongregarious and is also generally small in size. More species of fish have been identified from the Amazonian drainage system than from the Congo and Mississippi combined, encompassing not only a fascinating array of sizes, forms, and colors, but exhibiting special physiological and behavioral adaptations to the high acidity, nutritional deficiency, and other unique characteristics of their aquatic habitat. To label Amazonia an ecosystem of fantastic complexity, infinite diversity, and marvelous integration is barely to do justice to this masterpiece of natural selection. Its complexity, diversity, and integration are not fortuitous products of the evolutionary process, however, but crucial aspects of the configuration. For Amazonia with all its wonderful intricacy is like a castle built on sand. The foundation contributes nothing to the strength of the structure, and if enough components are removed or the bonds between them are sufficiently weakened, the entire configuration will collapse and disappear. This is not merely a theoretical judgment based on soil composition, rainfall, temperature, chemical and physical processes, and other constituent factors; it is a conclusion increasingly supported by observation of the effects of modern human exploitation.

Man has not always been a disruptive element in Amazonia. On the contrary, for millennia after his arrival he remained a harmonious member of the biotic community. The first immigrants were hunters and gatherers of wild foods who moved camp every few days, as the Síriono and Kayapó still do during the dry season. Each wandering band probably consisted of an extended family and this kin group remained the minimal social unit in aboriginal Amazonian society. As the lowlands became populated, wandering tended to become less random and increasingly confined within recognized territorial boundaries. Concomitantly, local variations in the annual subsistence round began to appear, taking advantage of regional differences in the kind, abundance, and seasonal availability of wild plants and animals. Some of the differences in the patterns of wild food utilization exhibited by subsequent agriculturalists undoubtedly stem from this early process of subsistence adaptation.

By 1000 B.C., if not earlier, domesticated plants had become an important subsistence component in the Amazonian lowlands. Dependence on agriculture both requires and permits a more sedentary way of life: requires because gardens must be planted, tended, and harvested; permits because food becomes available in greater concentration and local abundance. Since sedentary life is prerequisite to the accumulation of goods (which in turn makes possible occupational specialization, differential wealth, concentration of power, and many other technological, social and religious developments), and since it also offers the individual an improved chance of survival in the event of illness or infirmity, increasing sedentariness inevitably constitutes a primary tendency in the evolution of culture. In a habitat such as Amazonia, however, increasingly large and more permanent population concentrations conflict with the primary adaptive strategy, which emphasizes dispersal and transience. A compromise had to develop that provided the maximum benefits of settled life at the cost of minimal irreversible damage to the environment.

The pattern of culture that arose in Amazonia is almost as remarkable an adaptive configuration as the rain forest vegetation. Warfare provides excitement and a means of acquiring prestige, but also helps to prevent population increase. Sorcery not only explains the occurrence of death in a culture ignorant of germs and infection, but inhibits expansion of community size. Even a superficial examination of the cultures of the terra firme reveals numerous practices that are directly or indirectly adaptive; careful analysis would undoubtedly reveal more subtle examples of the interaction between culture and environment. The aboriginal adaptation to Amazonia is not only an ecological success, in the sense that it maintains a balance that is close to equilibrium, but it also provides to the human population a healthful and psychologically satisfying way of life.

The discovery of the Amazon by European explorers in the sixteenth century initiated a period of rapid and drastic change. New and lethal diseases decimated the indigenous population and alien cultural attitudes replaced those nurtured during millennia of natural selection. To foreign eyes, Amazonia was primarily a source of exotic products that could be sold for high prices, and the lure of immediate profit took precedence over the advantages of long-range productivity. The newcomers retained their traditional dietary preference for beef, rice, and coffee and continued to operate as an extension of European society, in which a highly diversified division of labor was fused with a complex system of commercial exchange. Since access to a market became the primary consideration, settlement concentrated on the riverbanks, leaving the hinterland inhabited only by scattered remnants of aboriginal tribes. Racial mixture created a biological blend between white, black, and Indian, but cultural integration was less successful. With a few major exceptions, such as house construction, European tastes and tools have prevailed over indigenous ones. Essential articles, such as clothing, hammocks, cooking utensils, knives, and axes, are available only by purchase. Since rubber, hides, Brazil nuts, and other kinds of commercially valuable forest products bring a low rate of return per man-hour expended, little time remains for subsistence activities. The inevitable result has been a decline in the quality of nutrition, which in turn lowers resistance to disease. Clearly, the post-European occupation of Amazonia has been an ecological disaster, initiating an accelerating incompatibility between the culture and the environment; it has also been a human disaster since it has condemned the Europeanized population to a desolate and hopeless existence in which physical survival is the overriding preoccupation.

Amelioration of this undesirable state of affairs will require a far better understanding than we now possess of the manner in which culture interacts with environment, but the effort to achieve it is hampered by serious analytic problems because the adaptive aspects of a custom or belief are not apparent to the members of the society in which it occurs. On the contrary, the people often provide convincing reasons for their behavior that have nothing to do with the environment. This being the case, it is probable that those elements and structures in our own culture fundamentally responsible for our nonadaptive behavior are masquerading under some other overt justification. Furthermore, the same general kind of behavior may have distinct adaptive and evolutionary functions in different contexts. Warfare is an example: in Amazonia it appears to be an important device for preventing population growth and concentration, but in other times and places it has served as a mechanism for the unification of independent political units, or for strengthening social stratification, or for replacement of one cultural configuration by another. To complicate the matter further, many traits simultaneously promote integration within the group and moderate the culture-environment articulation. Under such a circumstance, behavior that ceases to be adaptive may continue to provide vital psychological support to the population. Since loss of morale is a more immediate threat to survival than ecological disharmony, which usually manifests itself so slowly as to be imperceptible, natural selection will favor retention of such a trait until the ecological consequences reach a critical level. It seems probable that conflicts of this kind underlie the extinction of many once-flourishing cultural configurations and that they are a contributing factor in our own ecological crisis.

CULTURE AS A FORM OF BEHAVIORAL ADAPTATION

A major obstacle to progress in cultural ecology is the conviction that because man has evolved a unique kind of behavioral adaptation he is immune to the effects of natural laws. Yet it should be obvious to any unbiased observer that this planet got along very well without *Homo sapiens* for a remarkably long time. Life has existed for more than three billion years, terrestrial plants for less than half a billion years, mammals for about 200 million years, *Homo sapiens* for less than one million years, "civilization" only a few thousand years. Man is the product of a process that began eons before he appeared and is complex beyond his comprehension. He survived initially because of the adaptiveness of his biological makeup; he enhanced his adaptability by the acquisition of culture. Cultural adaptation has a distinct advantage over biological adaptation for a complex animal with a relatively long life span and a low reproductive capacity. Although hundreds of generations are required for major genetic alterations, drastic cultural changes can be made literally overnight. The fact that natural selection is provided with a new medium on which to operate, however, does not imply that the rules of the game and the objective have significantly changed. On the contrary, the similarity in the behavior of biological and cultural phenomena indicates that the same processes underlie both cultural and organic evolution.

A tendency toward diversification is one of the outstanding characteristics of both biological and cultural phenomena. On the organic level, the advantages of variety are obvious. Organisms that differ in their food preferences, reproductive behavior, and other habits avoid direct competition with one another and thereby improve their chances of survival. Each population exploits its own niche, and the more niches that are exploited the more efficiently the habitat is utilized. Differentiation has another aspect, however, which is of far greater evolutionary significance. Natural selection operates to bring existing populations into a higher state of adaptiveness to existing conditions; it cannot foresee future conditions and consequently cannot prepare for them. The best way to assure survival under such circumstances is to produce so many kinds of configurations that, regardless of what happens in the future, a few will possess the requisite adaptive properties. In other words, they will be "preadapted," not through foresight but as an accidental by-product of adaptation to their previous environmental niche.

Cultural diversity is amply attested in both the archeological and ethnographic records, and it seems clear that it has the same explanations as biological variation: namely, it allows more effective exploitation of existing habitats and provides a maximum number of potential pathways to the future. The first of these aspects is exemplified in the aboriginal adaptation to the Amazonian terra firme; the second permitted the evolution of civilization when circumstances were favorable.

NATURAL SELECTION AND THE DEVELOPMENT OF URBAN CIVILIZATION

The correlation between cultural diversity and evolutionary potential for cultural advance can be clarified by examining the appearance of urbanism. This innovation was as significant a breakthrough for cultural evolution as was the appearance of terrestrial life for biological evolution. Both initiated an accelerating process of diversification and increasing complexity that is unlikely to have reached its culmination even in the fantastic array of incredibly intricate organisms and cultures that fill the biosphere today. Biologists attribute this advance to natural selection, which favored the survival and amplification of characteristics that facilitated adaptation to a multitude of environmental niches. Anthropologists, on the other hand, have attempted to explain the origin of urban civilization by comparing the settlement pattern, technology, sociopolitical organization, and religious characteristics of early civilizations in Mesoamerica and Mesopotamia (where the archeological records are most complete), and trying to identify regularities in the configurations that emerged in these two areas. The environment is generally viewed either as a constant or as a pliant substrate, the object of a variety of cultural manipulations or exploitations rather than an active component of the evolutionary process.

If culture is basically a specialized means of adaptation employed by one species of mammal, however, then it follows that natural selection must have played as significant a role in cultural evolution as it has in biological evolution. To test this assumption requires us to establish two facts: 1) that city life is more adaptive (that is, it has more survival potential) than village life for a community; and 2) that the appearance of urbanism is correlated with a significant change in selective pressures. Even a casual examination of the evidence indicates that both of these propositions are correct.

Urban civilization has a number of characteristics that make it superior to tribal society. Its higher population size and density not only provide a stronger buffer against depletion from disease, famine, massacre, or other catastrophe, but also make possible a larger amount of internal cultural variation. More importantly, however, it represents a new kind of internal integration offering new adaptive opportunities. Like organisms, cultures cannot exceed a certain size without improving their internal organization. A city is not simply an enlarged village or a sedentary band, any more than a man is an inflated amoeba. A city differs from a village in the same ways that a man differs from an amoeba: in increased internal differentiation and in a higher level of integration, in which some segments exercise dominance over others. The same kinds of selective pressures underlie the evolutionary trend toward increasing size and complexity both in animals and in cultures. Under most circumstances, a large animal or a large community has access to a more sizable subsistence area than a small one; it also has an advantage over competitors and predators (until they also increase in size). Without improved integration between the parts, however, increase in size would be accompanied by loss of operational efficiency. Also, some kind of internal hierarchy is required (in other words, someone has to be in charge). In organisms, this has led to the development of a central nervous system; in cultures, it has produced a ruling class. Since this tighter integration improves adaptation, it is favored by selection via a complicated network of correcting and intensifying feedback reactions: increased food supply→ increased population density→ differentiation of function→ increased efficiency resulting from specialization→ improved food supply→ increased density→ improved mechanisms for internal exchange of goods and services→ specialized methods of defense; and onward and upward until the environmental potential is fully exploited.

If urbanism is so highly adaptive, why has it not appeared everywhere in the world? The answer should lie in the kinds of selective pressures that are exerted by different kinds of environments. We have noted the incipient expression of occupational division of labor, social stratification, and other characteristics of urban society among several of the Amazonian terra firme groups. Analyzing these in the context of the total environment, we concluded that they were simply local variations of a generalized pattern of tropical forest culture with no special adaptive value. Suppose, however, that this same range of sociopolitical diversity existed among groups inhabiting a different type of environment. To be specific, suppose that this environment were Mesoamerica, where the subsistence limitations characteristic of the Amazonian terra firme do not exist. Under more propitious conditions, selection would be expected to favor and thus gradually to intensify features compatible with increasing population density. Any group that possessed such traits in incipient form would have a selective advantage and would tend to increase its density, or expand its geographical range, or both, at the expense of neighboring groups.

The archeological record in Mesoamerica is in fact characterized by successive attempts at domination of increasingly larger territories by competing "states." Furthermore, these competitors appear to have possessed different types of integrating mechanisms, just as would be expected if they represented evolutionary elaborations of cultures with different combinations of incipient features. Some emphasized production and redistribution of goods through trade and markets; some intensified religious concepts and practices; a few tried to modify cooperative patterns of kinship behavior to satisfy new requirements; others substituted social segments based on occupation, residence, wealth, and other criteria for the former kin groupings. As would be expected by analogy with biological evolution, not all these variations were equally successful in the long run. The biological model of adaptive diversification explains both the kind of cultural variability that exists in Amazonia and the general process by which civilization evolved in Mesoamerica, by showing them to be expressions of a single adaptive process.

THE ADAPTIVE SIGNIFICANCE OF CULTURAL ISOLATING MECHANISMS

The point has often been made that culture is learned behavior and consequently is more freely disseminated through time and space than are biological characteristics, which cannot cross genetic barriers. While some noncultural behavior can be learned, a marked difference clearly exists in the potential diffusibility of biological and cultural innovations. On the other hand, the limited distribution of many cultural features and the existence of pronounced disparities between geographically propinquitous groups make it evident that the capacity for dissemination inherent in culture is seldom realized. The persistence of distinctions between cultures in frequent communication implies that the latent diffusibility of culture is denied free expression, and anyone familiar with biological evolutionary theory will immediately suspect that isolating mechanisms must exist. Among plants and animals, where exchange of characters occurs through interbreeding, isolation is accomplished by a change in behavior that reduces the frequency of mating between two populations, allowing them to diversify genetically until they become so different that they are no longer able to produce viable offspring. The process is gradual and may not culminate in biological sterility, so that some gene flow can continue via occasional individuals who happen to interbreed. Behavioral isolation thus opens the door to diversification (with its important adaptive advantages) without completely shutting off the flow of new genes from closely related races or subspecies.

From an evolutionary standpoint, the problem of balancing the benefits of diffusion against the advantages of diversification is the same for cultural as for biological phenomena. Since culture is potentially more easily transmitted than genetic variation, however, cultural diffusion has a much greater capacity not only to prevent differentiation from occurring but also to inject new and incompatible traits into previously isolated complexes and thus to destroy them. The breakdown of primitive societies throughout the world in recent decades under the impact of acculturation, with traumatic effects on the populations involved, is a vivid demonstration of the destructive potential of uncontrolled cultural dissemination and of the importance of isolating mechanisms for the development and maintenance of adaptive cultural configurations.

Since culture is learned, it is not surprising to find that the principal barriers to its transmission also are learned. One of the most effective, to judge from its widespread occurrence, is ethnocentrism, or the conviction that one's own people are "true men" while all other groups are inferior, if not subhuman. The behavior of such inferiors is not only considered unworthy of imitation, but may even be viewed as inalienable along with their hair color and other biological traits. This psychological attitude remains widespread today, and is expressed in contempt for the food habits, costume, methods of personal adornment, laziness, aggressiveness, or some other characteristic of another group and a consequent effort to avoid association with such individuals as much as possible, even to the extent of sacrificing personal comfort or material advantage. The validity of such biases does not effect their utility as isolating mechanisms. In spite of their seeming superficiality, they have helped to make cultural evolution possible and continue to promote diversification in the interests of the survival of mankind as a whole.

Among the terra firme tribes of Amazonia, another important barrier to cultural exchange is provided by supernatural beliefs. The role of sorcery in population density control was mentioned earlier. In addition, fear of sorcery is an important isolating mechanism since it operates most strongly between adjacent groups, which tend to be most similar and consequently most susceptible to mutual influence. Furthermore, the fact that sorcery is more successful if it can be practiced on an object obtained from the victim reduces transfer of material goods to a minimum. Another supernatural concept that inhibits interaction and consequently reduces the rate of cultural diffusion is the notion that the territory outside the tribal boundary is infested with hostile spirits. Like biological isolating mechanisms, such attitudes promote in-group cohesion while simultaneously minimizing between-group relations.

Once a well differentiated and highly adapted cultural configuration has emerged, it becomes resistant to the intrusion of alien traits, just as a species becomes impervious to genetic disruption after a sterility barrier has developed. One of the best examples of cultural impenetrability can be observed along the Amazonian-Andean interface. At the time of the Spanish conquest, most of the Andean area was dominated by the Inca Empire, which had been formed only a few decades earlier by progressive military conquest. At its culmination, it extended from northern Ecuador to central Chile and incorporated a population of several million persons. The administration of this elongated territory required rapid communication, and an ingenious system of roads and human couriers was established for the purpose. Since the speed with which news of rebellions, earthquakes, or other disruptive incidents reached the capital in Cuzco was proportional to the distance overland, control could never be as effective toward the northern and southern margins of the empire as in the central part. Clearly, a less attenuated territory of equivalent area would have been a more efficient administrative unit, and departure from this ideal form implies some overriding factor in the local situation. On the basis of evolutionary theory, we can hypothesize that this factor was ecological and that the cost of maintaining the Inca type of society in the adjacent Amazonian forest would have been higher than the cost of preserving a far-flung integration in a more uniform environmental setting. The validity of this inference is borne out by ethnohistorical documents, which record repeated unsuccessful military efforts to incorporate the eastern lowlands into the Inca Empire.

In spite of the fact, attested by archeological evidence, that the Andean and Amazonian peoples engaged in trade from the time they began settled agricultural life (if not earlier), this continuous communication failed to prevent the emergence of two distinct types of cultural configurations. With the passage of time, selection favored those characteristics best suited to the exploitation of each environment, with the result that objects, beliefs, and practices suited to one region were increasingly likely to be useless in the other. This dichotomy is reflected in the character of the trade items found in coastal archeological sites, which consist of raw materials, such as feathers, hides, wood, fruits, and drugs, rather than finished objects. Long before the appearance of the Inca Empire, two distinct configurations had emerged, each of which was supreme in its own habitat but neither of which could make any significant impact on the territory of the other. The ecological barrier is sufficiently strong that modern industrial society, which replaced the Inca in the highlands, has been unable to implant itself in the tropical rain forest environment.

EVOLUTION AS A UNIVERSAL PROCESS

Similar parallels between cultural and biological phenomena could be multiplied indefinitely, and for a very good reason. The processes involved are not biological but universal and clearly underlie all change regardless of whether physical, biological, or cultural phenomena are involved. Because of the historical circumstance that Darwin's *On the Origin of Species* had a more profound impact than contemporary attempts by Spencer, Tylor, and others to reconstruct the evolution of human society, the impression has emerged that resemblances between the behavior of cultural and biological phenomena are simply analogies—and misleading ones, at that. This bias has prevented the application of numerous insights derived from biological research to the explanation of cultural adaptation and evolution.

The recognition that evolution is a universal process and that diversification and natural selection operate with equal force on biological and cultural phenomena does not imply that all the types of interactions responsible for biological adaptation are equally significant mechanisms of cultural change. Just as plants differ from animals in mobility, longevity, method of reproduction, and other significant features, so cultures differ from biological organisms. Certain processes of adaptation are more common in animal than in plant populations, and we should expect to find equivalent differences in emphasis between cultural and biological phenomena.

While biological evolutionary theory can provide valuable leads for cultural analysis, this interdisciplinary cooperation is not a one-way relationship. Biologists are limited to observing and measuring the organisms they study, whereas students of human behavior have access to the psychological dimension of adaptation. We can investigate the way in which attitudes, feelings, beliefs, and ideas interact with behavior, and consequently we can expose their role in the development and preservation of behavioral differences. Thus far, our own psychological involvement has handicapped achievement of the objectivity needed to see culture as the product of natural selection rather than human ingenuity. If this obstacle can be overcome, understanding of the complicated manner in which psychological factors are involved in adaptation, and thus contribute to the evolutionary process, will surely emerge as a new and exciting field for scientific exploration.

Selected References

This listing has been restricted to references quoted or utilized extensively; many have bibliographies that can be consulted for further reading. All quotations not originally in English have been translated and all metric measurements have been converted and rounded off.

Acuña, Cristóbal de

1942 *Nuevo descubrimiento del gran Río de las Amazonas.* Buenos Aires, Emecé Editores.

Ashburn, P.M.

1947 *The ranks of death; a medical history of the Conquest.* New York, Coward-McCann.

Banner, Horace

1961 *O índio Kayapó em seu acampamento.* Belém, Boletim do Museu Paraense Emílio Goeldi, Antropologia 13.

Bitencourt, Agnelo

1950–51 "Aspectos da pesca na Amazônia." *Boletim da Sociedade Brasileira de Geografia* 1:137–144.

Carvajal, Gaspar de

1934 *The discovery of the Amazon, according to the account of Friar Gaspar de Carvajal and other documents.* Compiled by José Toribio Medina, edited by H. C. Heaton. Special Publication 17. New York, American Geographical Society.

Cruz, Laureano de la

1900 *Nuevo descubrimiento del río de Marañon llamado de las Amazonas.* Madrid, Bibl. de La Irradiación.

Danielssen, Bengt

1949 "Some attraction and repulsion patterns among Jibaro Indians; a study in sociometric anthropology." *Sociometry* 12:83–105.

Dreyfus, Simone
1963 *Les Kayapo du nord; état de Para—Brésil: contribution à l'étude des Indiens Gé.* Paris, Ecole Pratique des Hautes Etudes, Sorbonne.

Fock, Niels
1963 *Waiwai; religion and society of an Amazonian tribe.* Copenhagen, Nationalmuseets Skrifter, Etnografisk Roekke 8.

Frikel, Protásio
1959 *Agricultura dos índios Mundurukú.* Belém, Boletim do Museu Paraense Emílio Goeldi, Antropologia 4.
1968 *Os Xikrín; equipamento e técnicas de subsistência.* Pubs. Avulsas 7. Belém, Museu Paraense Emílio Goeldi.

Fritz, Samuel
1922 *Journal of the travels and labours of Father Samuel Fritz in the river of the Amazons between 1686 and 1723.* George Edmundson, Ed. 2nd. series, No. 51. London, Hakluyt Society.

Gibbs, Ronald J.
1967 "The geochemistry of the Amazon river system. Part I, The factors that control the salinity and the composition and concentration of the suspended solids." *Geological Society of America Bulletin* 78:1203–1232.

Gonçalves, Augusto Cezar Lopes
1904 *The Amazon; historical, chorographical and statistical outline up to the year 1903.* New York, H. J. Hanf.

Gourou, Pierre
1966 *The tropical world; its social and economic conditions and its future status.* New York, Wiley.

Guerra, Antonio Teixeira, Editor
1959 *Geografia do Brasil; Grande região norte.* Biblioteca Geográfica Brasileira, Vol. 1, Publ. 15. Rio de Janeiro, Conselho Nacional de Geografia.

Hegen, Edmund E.
1967 "Man and the tropical environment; problems of resource use and conservation." *Atas do Simpósio sôbre a Biota Amazônica* 7:165–175.

Heriarte, Maurício de
1874 *Descripção do Estado do Maranhão, Pará, Corupá e Rio das Amazonas.* Vienna, Imprensa do filho de C. Gerold.

Holmberg, Allan R.
1950 *Nomads of the long bow; the Sirionó of eastern Bolivia.* Publ. 10. Washington, D.C., Institute of Social Anthropology.

Instituto Brasileiro de Geografia e Estatística
1966 *Atlas nacional do Brasil.* Rio de Janeiro.
Karsten, Rafael
1935 *The head-hunters of western Amazonas; the life and culture of the Jibaro Indians of eastern Ecuador and Peru.* Commentationes Humanarum Litterarum, Tomus VII, Nr. 1. Helsingfors, Societas Scientarum Fennica.
Ladell, William S.S.
1964 "Terrestrial animals in humid heat: man." Dill, D.B., E.F. Adolf and C.G. Wilber, Eds., *Handbook of Physiology,* Section 4:625–659. Washington, D.C., American Physiological Society.
Lima, Rubens Rodrigues
1956 *A agricultura nas várzeas do estuário do Amazonas.* Belém, Boletim Técnico do Instituto Agronómico do Norte 33.
Mendes, Jose Amando
1938 *As pescarias amazônicas e a piscicultura no Brasil.* São Paulo, Livraria Editora Record.
Mohr, E.C.J. and F.A. Van Baren
1954 *Tropical soils; a critical study of soil genesis as related to climate, rock and vegetation.* London, Interscience Publishers.
Nimuendajú, Curt
1949 "Os Tapajó." Boletim do Museu Paraense Emílio Goeldi 10:93–106.
Nye, P.H. and D.J. Greenland
1960 *The soil under shifting cultivation.* Technical Communication 51. Harpenden, Commonwealth Bureau of Soils.
Oberg, Kalervo
1953 *Indian tribes of northern Mato Grosso, Brazil.* Publ. 15. Washington, D.C., Institute of Social Anthropology.
Raleigh, Walter
1848 *The discovery of the large, rich, and beautiful empire of Guiana . . . performed in the year 1595.* R.H. Schomburgk, Ed. London, Hakluyt Society.
Ribeiro, Darcy
1967 "Indigenous cultures and languages of Brazil." Janice Hopper, Ed., *Indians of Brazil in the Twentieth Century,* pp. 77–165. Washington, D.C., Institute for Cross-Cultural Research.
Richards, Paul W.
1952 *The tropical rain forest.* Cambridge, Cambridge University Press.

Setzer, José
1967 "Impossibilidade do uso racional do solo no alto Xingú, Mato Grosso." *Revista Brasileira de Geografia* 29:102–109.

Simon, Pedro
1861 *The expedition of Pedro de Ursua and Lope de Aguirre in search of El Dorado and Omagua in 1560–61.* 1st Series, No. 28. London, Hakluyt Society.

Simpson, George Gaylord
1953 *The major features of evolution.* New York, Simon and Schuster.

Sioli, Harald
1964 "General features of the limnology of Amazônia." *Verh. Internat. Verein. Limnol.* 15:1053–1058.

Soares, Lúcio de Castro
1953 "Limites meridionais e orientais da área de ocorrência da floresta amazônica em território brasileiro." *Revista Brasileira de Geografia* 15:3–122.

Yde, Jens
1965 *Material culture of the Waiwai.* Copenhagen, Nationalmuseets Skrifter, Etnografisk, Roekke 10.

Suggested Further Reading

Boughey, Arthur S.
1968 *Ecology of populations.* [The relationship between organisms and the environment, including population density, speciation, food chains, etc.] New York, Macmillan.

Chagnon, Napoleon A.
1968 *Yanomamö; the fierce people.* [A large unacculturated group living on the Brazil-Venezuela border.] New York, Holt, Rinehart and Winston.

Goldman, Irving
1963 *The Cubeo; Indians of the northwest Amazon.* [A tribe of the Rio Uaupes, on the Colombia-Brazil boundary.] Urbana, University of Illinois Press.

Murphy, Robert F. and Buell Quain
1955 *The Trumaí Indians of central Brazil.* [The life of now extinct neighbors of the Camayurá in the upper Xingú.] New York, American Ethnological Society.

Shepard, Paul and Daniel McKinley, Eds.
1971 *Environ/mental.* [Twenty chapters dealing with ecological implications and consequences of human intelligence.] Boston, Houghton Mifflin.

Vayda, Andrew P., Ed.
1969 *Environment and cultural behavior; ecological studies in cultural anthropology.* [The adaptive significance of warfare, ritual practices, family organization, divination, psychological and physical stress, and other kinds of behavior.] Garden City, N.Y., The Natural History Press.

Wagley, Charles
1953 *Amazon town; a study of man in the tropics.* [Rural life in Amazonia today.] New York, Macmillan.

modern inhabitants to have dangerous magical powers, particularly in the area of sex and love.

Frutão (*Pouteria pariri*). A large green-skinned fruit with a sticky fibrous cream-colored pulp, which is eaten raw.

Ité (*Mauritia flexuosa*). A palm with a nutritious fruit with an orange-yellow pulp; the unopened leaves are widely used for cordage.

League. An unstandardized measure of distance, varying from about 2.4 to 4.6 miles.

Maize (*Zea mays*). The principal New World cereal, commonly known as corn.

Manatee (*Trichechus inunguis*). Aquatic mammal, resembling a seal but larger, which subsists on the succulent grasses of the várzea lakes. The hide is thick and extremely tough; the meat is excellent.

Mandive. Synonym for bitter manioc.

Mangaba (*Harcornia speciosa*). A savanna tree about 15 feet tall that bears a spherical yellow fruit resembling an apricot. The soft whitish pulp has a slightly acid, cherry-like flavor.

Manioc (*Manihot esculenta*). The staple root crop throughout the tropical lowlands of America. Although botanists recognize a single domesticated species, cultivars vary greatly in the amount of cyanogenetic glucosides they contain. Those with low concentrations are known as "sweet" and the tubers are edible when boiled; those with high concentrations are "bitter" and must be processed to remove the poisonous juice before the tuber can be safely eaten.

Matamatá (*Chelys fimbriata*). An ugly long-necked turtle found in muddy places and stagnant water; delicious in spite of its unappetizing appearance.

Motacú (*Scheelea princeps*). A palm with an edible fruit.

Mussuan (*Cinosternum scorpioides*). A small land turtle, comparable in size to the eastern United States box turtle; the meat is tasty but sparse.

Paca (*Coelogenys paca*). A short-legged tailless rodent, largely nocturnal. In pattern and color, its hide resembles that of a fawn.

Palmito. The terminal shoot of the growing palm, which can be eaten raw or cooked. While most species are edible, some have a bitter taste.

Papaya (*Carica papaya*). A cultivated fruit tree. The fruit, which grows from the trunk at the base of the leaf crown, is often more than a foot long; the flesh is light to bright red-orange under the thin green or yellow skin and has a sweet and delicate flavor.

Peach palm. Synonym for chonta.

Peccary (*Dicotyles labiatus, D. torquatus*). Two species of wild pig, both of which travel in large and noisy bands (occasionally exceeding 100 individuals) through the forest in search of food. The flesh tastes like pork.

Pescada (*Plagioscion squamosissimus*). An abundant and flavorful scaled fish.

Phytoplankton. Plant organisms, often microscopic in size, floating in lakes and ponds.

Piqui (*Caryocar* sp.). A tall (50 feet) tree bearing spherical fruits about the size of a large orange. Each fruit contains 1 to 4 segments composed of a kidney-shaped seed covered with a yellowish oily pulp.

Piraíba (*Brachyplatystoma* sp.). The largest Amazonian catfish.

Pirarucú (*Arapaima gigas*). The largest scaled fish in the Amazon, delicious fresh or dried and salted.

Rebec. A lute-shaped musical instrument used in medieval Europe.

Samuque (*Syagrus botryophora*). A palm with an edible seed.

Shaman. A person with special connections with supernatural forces or beings, which give him the power to cure illness, practice sorcery, and perform other similar deeds. Also known as a medicine man.

Sweet potato (*Ipomoea batatas*). A staple root crop growing throughout lowland tropical America.

Tambaqui (*Myletes bidens, Colossoma bidens*). A fruit-eating fish especially common in August and September in lakes and rivers; caught with hook or net.

Tapir (*Tapirus americanus*). The largest Amazonian terrestrial mammal, primarily nocturnal, and a good swimmer.

Tartaruga (*Podocnemis expansa*). The largest water turtle, found in the Amazon and its major tributaries below the first rapids; captured on beaches by overturning and in the water by shooting or harpooning. The eggs (which resemble ping-pong balls) have a soft papery shell and consist mainly of yoke.

Terra firme. Land not subject to annual inundation; elevation varies from immediately above flood level to several thousand feet.

Toucan (several species). A large-billed bird with red and yellow breast feathers, which are used in making ornaments.

Tracajá (*Emys tracaja*). A water turtle, smaller, less abundant, and more widespread than the tartaruga; both eggs and adults are considered superior in flavor to those of the tartaruga.

Tucunaré (*Cichla ocellaris*). A large fish prized both for its flavor and because it has few bones.

Várzea. The flood plain of a white water river, which receives an annual deposit of fertile silt.

Yam (*Dioscorea* sp.). The common domesticated yams are Old World plants introduced into Amazonia after European contact. A few native species were used indigenously, especially in eastern Brazil.

Yuca, yucca. Synonym for sweet manioc.

Zapote (*Lucuma obovata*). A fruit tree of eastern Ecuador, related to but not the same as the Mexican zapote.

Index